Robert Saundby

Lectures on Bright's Disease

Robert Saundby

Lectures on Bright's Disease

ISBN/EAN: 9783744718172

Printed in Europe, USA, Canada, Australia, Japan

Cover: Foto ©berggeist007 / pixelio.de

More available books at **www.hansebooks.com**

LECTURES

ON

BRIGHT'S DISEASE.

BY

ROBERT SAUNDBY, M.D. Edin.,

FELLOW OF THE ROYAL COLLEGE OF PHYSICIANS, LONDON; EMERITUS SENIOR PRESIDENT
OF THE ROYAL MEDICAL SOCIETY; FELLOW OF THE ROYAL MEDICO-CHIRURGICAL
SOCIETY; MEMBER OF THE PATHOLOGICAL SOCIETY OF LONDON; PHYSICIAN
TO THE GENERAL HOSPITAL; CONSULTING PHYSICIAN TO THE EYE
HOSPITAL; AND CONSULTING PHYSICIAN TO THE HOSPITAL
FOR DISEASES OF WOMEN, BIRMINGHAM.

WITH FIFTY ILLUSTRATIONS.

BRISTOL: JOHN WRIGHT & CO.
LONDON: HAMILTON, ADAMS & CO., PATERNOSTER ROW.
1889.

PREFACE

IN presenting this work for the indulgent consideration of my professional readers, I would disclaim all pretension to have said the last word on the many vexed questions with which its subject matter is encumbered. I have endeavoured to explain within a modest compass, the present state of contemporary knowledge, making such additions and suggestions as have resulted from thirteen years' clinical and pathological study of *Bright's Disease*. Some of the material for this book has been previously published, but every chapter has been re-written, every point has been thought out again, and I need offer no apology for such alterations of opinion as may be perceptible in these pages.

Except where the source is acknowledged, all the figures have been drawn by myself from preparations in my possession. I desire to express my indebtedness to Dr. G. F. CROOKE, Pathologist to the General Hospital, for many beautiful microscopical preparations; to Dr. MACMUNN, and Mr. EALES, for kindly overlooking portions of the proof-sheets; to the publishers of LANDOIS and STIRLING's Physiology for permission to use certain illustrations of the Anatomy of the Kidney; and to all preceding writers on Urinary Disease, among whom I desire to mention SIR WILLIAM ROBERTS, whose work on Urinary and Renal Diseases must remain for all time a never-failing source of sound clinical information.

R. S.

BIRMINGHAM,
    February, 1889.

CONTENTS.

Section I.—GENERAL PATHOLOGY.

CHAP.		PAGES
I.	ALBUMINURIA	1—22
II.	THE PATHOLOGY OF DROPSY	23—36
III.	PATHOLOGICAL RELATIONS OF TUBE CASTS	37—44
IV.	CARDIO-VASCULAR CHANGES	45—63
V.	PATHOLOGY OF POLYURIA	64—68
VI.	PATHOLOGY OF URÆMIA	69—80
VII.	RETINAL CHANGES	81—95

Section II.—CLINICAL EXAMINATION OF THE URINE.

VIII.	THE URINE IN HEALTH AND DISEASE	97—146

Section III.—BRIGHT'S DISEASE.

IX.	HISTORY—CLASSIFICATION—ETIOLOGY	147—163
X.	GENERAL ANATOMY OF THE KIDNEY	164—171
XI.	FEBRILE NEPHRITIS	172—198
XII.	LITHÆMIC NEPHRITIS	199—230
XIII.	OBSTRUCTIVE NEPHRITIS	231—242
XIV.	COMPLICATIONS OF CHRONIC BRIGHT'S DISEASE	243—268
XV.	TREATMENT OF LITHÆMIC NEPHRITIS	269—282
	GENERAL INDEX	283—290

LECTURES ON BRIGHT'S DISEASE.

Section I.—GENERAL PATHOLOGY.

ALBUMINURIA.

BY ALBUMINURIA we mean the presence in the urine of an albuminous body which is coagulated by heat or precipitated by neutralisation.

Serum albumen is the substance ordinarily found in albuminous urine. It is soluble in water, coagulable by heat at from 73° to 75° C., and precipitated by acids; it is readily soluble in strong nitric acid, and not precipitated by common salt, carbonates of the alkalies or very dilute acids.

Syntonin, or acid albumen, is formed from serum albumen in the presence of a free acid; *alkali-albumen* is a similar modification caused by the presence of a free alkali; these bodies are not coagulated by heat, but are precipitated by neutralisation.

Syntonin may be formed by the careless use of a test tube containing traces of nitric acid.

Alkali-albumen is often present in putrid urine.

Serum globulin is insoluble in distilled water, but dissolves readily in the presence of common salt, and is then coagulable like serum albumen at 73° to 75° C.

It may be precipitated from its solutions by carbonic acid gas, or by saturation with common salt or magnesium sulphate. It differs moreover from serum albumen in its more ready diffusibility through animal membranes. It is always present in albuminous urine. In the blood the proportion of serum globulin to serum albumen is as 2 to 3; but in urine, owing to its higher rate of diffusion, this proportion may be reversed. It has been more than once suggested that it is the body present in functional albuminuria or after paroxysmal hæmoglobinuria (GULL, RALFE); but this does not appear to be the case. Out of 16 cases of functional albuminuria examined for this purpose, the albuminous body was pure globulin in only one case, and on a second examination serum albumen was present with it. It has been noted to be in excess in cantharides poisoning, in chronic nephritis with waxy degeneration, in the early stage of scarlatinal nephritis, in the albuminuria associated with dyspepsia, in phthisis, &c. Werner found globulin only present in a case of acute nephritis, and attributed it (following SENATOR) to the destruction of the renal epithelium which contains seven or eight times more globulin than serum albumen. Hermann found globulin to be chiefly present in a case of eclampsia of pregnancy with albuminuria.

Retractile albumen (BOUCHARD) was the name given to albumen which, on boiling, falls down in dense flakes; it was held to be characteristic of Bright's disease. It is worth mentioning in order to note that this has been abundantly proved to be a mistake, this peculiar behaviour being dependent upon the *acidity* of the urine (LÉPINE.)

Peptone is best known to us as the product of the digestive action of the gastric or pancreatic juices on albuminous food; but of late years it has been observed to be present in the urine in many acute and chronic diseases, generally in association with serum albuminuria. It must not be assumed that it is necessarily food peptone excreted by the kidneys. Peptone after absorption rapidly disappears, and seems to act as a poison if injected into the circulation; while it is said on good authority (GRÜTZNER) that digestive ferments are present in the urine. According to Mya and Belfanti, trypsin, the normal proteolytic ferment of the urine, is replaced by pepsin in Bright's disease; but either would be capable of effecting the conversion of albumen into peptone,—a process which is probably favoured by elevation of the body temperature in fever, and by the retention of the urine for a certain time in the bladder. Peptones are not coagulable by heat or precipitated by neutralisation; they are thrown down by several of the tests recommended of late years for albumen, *e.g.*, picric acid, metaphosphoric acid, Tanret's test (potassio-mercuric iodide), &c., but the characteristic reaction is a pink or rose violet colour with Fehling's solution in the cold.

Albumen in the process of conversion into peptone becomes in the first instance a substance similar to or identical with syntonin, and subsequently it is changed into a body called *propeptone* or hemi-albumose before its final conversion into peptone. This propeptone or hemi-albumose is nearly allied to the substance known as Bence Jones's albumen or par-albumen. It is probable that there are several closely related bodies forming a group, as the reactions given differ in various details. Propeptone, according to Salkowski, remains clear on boiling; it is precipitated on adding acetic acid and common salt solution, but clears up on heating, to be thrown down again on cooling. With pure nitric acid it is precipitated; on heating it dissolves, with the development of an intense yellow colour, and on cooling comes down again. Treated with caustic soda and cautiously with copper sulphate, it gives a purple violet which is lost if excess of copper sulphate is used. With phosphoro-tungstic acid, tannic acid, acetic acid and ferro-cyanide of potassium it is precipitated; but the last reaction does not occur in the presence of ammonia. Heated with a drop of Millon's reagent (acid nitrate of mercury), it gives a deep red colour and precipitate, but this reaction fails when much common salt is present.

Bence Jones described the albuminous body in his case as affording the following reactions; he wrote, "The urine that contained it did not give a precipitate immediately by nitric acid, and when heated it did not coagulate, and nitric acid when added to the boiling urine did not give a precipitate. If, after boiling, the urine was cooled, then the precipitate fell, but was immediately redissolved by heat." The difference between this and the reactions of propeptone given above is that it was not precipitated by nitric acid. Thormählen has described a form of albumen which differed from propeptone in the precipitate caused by nitric acid not redissolving on heating, and by the solution when cleared by heating, remaining clear on cooling. The urine of Joseph B——, a patient in the General Hospital, gave a similar reaction on one occasion.

In the urine of a case of pernicious anæmia I found an albuminous body which presented most of the reactions of propeptone, but on boiling there was a dense cloud which on adding acetic acid disappeared, while the whole fluid became gelatinised, so that the test tube could be turned upside down. Nitric acid gave a precipitate, but it dissolved on heating without any yellow coloration. Acetic acid and ferro-cyanide of potassium gave no reaction, but this may have been due to the presence of salts of ammonia.

Posner has found propeptone together with traces of peptone in human semen, as well as in the urine of spermatorrhœa and after pollutions. It has been noticed in many diseases, mollities ossium (BENCE JONES) syphilitic cachexia, diphtheria, cancer of œsophagus, muscular atrophy (SENATOR), petroleum inunction (LASSAR), chronic nephritis (HOPPE-SEYLER), and therefore has no special clinical significance.

Albuminuria once regarded as diagnostic of Bright's disease has lost this primitive meaning.

The number and variety of the pathological relations under which albumen may appear in the urine compel us to regard it as dependent not only upon inflammation, grave congestions and other coarse organic changes, but upon slight variations in the mechanical conditions of the circulation in the kidney. Excluding accidental admixtures of blood or pus from the bladder or urethra, albuminuria is met with not only in acute and chronic Bright's disease, but in diseases of the heart, lungs, and liver, in peritonitis, pregnancy, abdominal tumours, in most febrile and inflammatory diseases, in many cases of poisoning, in cancer, tubercle and syphilis, in lardaceous disease, in anæmia, debility, dyspepsia, purpura, scurvy, after paroxysmal hæmoglobinuria, in gout, in delirium tremens, in various diseases of the brain and spinal cord, in epilepsy, in certain skin diseases, as well as in apparently healthy persons after bathing, exercise, etc.

All these states are capable of being arranged under the following groups.

1.—*Congestions of the kidney.*

Active or arterial congestion may result from a chill to the skin as in bathing (MAHOMED, JOHNSON); from exposure to cold; from elimination of some irritant through the kidneys such as alcohol, uric acid, phosphorus, lead, cantharides, &c.; from the direct action on the kidneys of a morbid poison derived from the blood, such as the virus of scarlatina, diphtheria, typhoid, &c., a condition very liable to pass on to acute nephritis; or finally from vaso-motor paralysis after injuries to the spinal cord, and in some other nervous affections.

Passive or venous congestion may result from cardiac, pulmonary or liver disease, peritonitis, pregnancy. abdominal tumours, the hypostatic congestion of prolonged illness, failure of the circulation from enfeebled heart, in fever, in anæmia, in exophthalmic goitre, and fatigue, or after violent exertion, *e.g.*, epilepsy.

2.—*Inflammation*, acute or chronic, in many inflammatory, zymotic and septic diseases, in gout, in chronic lead poisoning, &c.

3.—*New growths.* Cancer, tubercle or syphilitic deposits in the kidney.

4.—*Degenerations.* Lardaceous degeneration of the renal arterioles.

5.—*Alterations in the composition of the blood*, as in purpura, or scurvy, and after attacks of paroxysmal hæmoglobinuria.

Looked at in this way the difficulties which have beset the discussion of the significance of albuminuria melt away; this result is attained by the absolute surrender of the doctrine that albuminuria signifies Bright's disease, and the acceptance of the view that it is simply the admixture of albumen derived from the blood serum with the urine. It is a fact beyond dispute that albumen may be present in the urine of persons apparently in good health. It is even contended by some that there is a trace of albumen in the urine which is physiological. Posner found an albuminous body precipitated by ferro-cyanide of potassium in every one of seventy experiments on the urine of healthy persons, and Chateaubourg using potassio-mercuric iodide found albumen in 592 out of 701 examinations of urine in healthy persons. I do not care to contend that there is not some departure in these cases from the normal mechanism of the renal circulation; but it is certain that neither this lesion nor the loss of albumen gives rise to any derangement of health which impairs the working capacity of the individual or tends to shorten his life; in other words, there are cases of albuminuria which not only do not require medical treatment, but may be safely accepted by life assurance companies. In these cases the urine is normal in every other respect, there are no tube casts, the amount of solid matter excreted is sufficient, and there are no signs of cardiac hypertrophy, of high arterial tension, no retinal changes and no œdema. Leube examined the morning urine of 119 healthy soldiers, of ninety once, of twenty-three twice, and of six three times. Albumen was found six times in the urine of five different soldiers, five times only a trace, once a distinct cloud. The latter urine was that of a soldier whose morning urine was examined twice and contained once a distinct cloud, the other time only a trace. The midday urine of 119 soldiers was examined, especially that passed after a five hours' march or many hours' parade, in the months of June, July and August. The five soldiers whose morning urine contained albumen also had albumen in their mid-day urine, three times a trace, three times more distinctly. Moreover, in 148 observations, albumen was found eighteen times in the urine of soldiers whose morning urine was quite free from albumen, in eight distinctly, in ten only a faint trace. The results were that 4·2 per cent. of the soldiers had albumen in their morning urine, and sixteen per cent. had albumen in their mid-day urine. No casts or blood corpuscles were found. All the cases in which albuminuria appeared were carefully examined, and urethral discharges, as well as signs of Bright's disease, were carefully noted to be absent. The test used was boiling and acetic acid; but the albuminous body was separated and tested by Millon's reagent, sulphate of copper, ferrocyanide of potassium, &c.

Grainger Stewart examined the urine of 505 presumably healthy individuals, comprising 205 soldiers of the Seaforth Highlanders, 100 healthy male adults in civil employment, 150 healthy inmates of Craiglockhart Poorhouse (100 adults, over sixty, and 50 children), and 50 children in the Orphan hospital; of these 166 or 32·8 had albuminuria. Of the 205 soldiers 77, or 37·56 per cent. had albuminuria; of 100 healthy male adults 10 only; of 100 inmates of Craiglockhart, about or above sixty years of age, 62 had albuminuria, and of 100 children it was present in 17. Dr. Stirling examined 461 healthy persons of whom 369 were boys, and of these 118 had albuminuria, of whom 77 were boys, giving for adults 44 per cent., for boys 20 per cent. In Leube's cases the albuminuria was found in the mid-day urine four times more frequently than in the morning urine, that is the urine passed immediately on rising. Stirling found the erect posture was the great determining factor in the production of albuminuria in his cases. The same fact is attested by Senator's experience, who found that the urine of himself and three clinical assistants at the Augusta hospital in Berlin repeatedly contained albumen between 11 a.m. and 12.30. Bull has published a case in which the albumen was always absent when the patient was in bed. Marcacci could produce albuminuria in himself by making rotatory movements with his arms for fifteen minutes. But this is trespassing on another part of the subject, and it is sufficient to state here that this form of albuminuria in healthy persons seems to depend essentially upon some mechanical condition connected with the erect position and exercise.

From his figures Grainger Stewart concludes that albuminuria is more common in health as life advances;

but this is doubtful. His elderly people show a high percentage, but there is no evidence that care was taken to eliminate prostatic and vesical catarrh or even latent Bright's disease,—all very common conditions at that age. Other observers have met with functional albuminuria most commonly in young men.

But there are other cases in which some departure from health is present, for example, atonic dyspepsia, with oxaluria or lithuria, anæmia perhaps not very marked; or the patient is an overgrown weakly person, with a tendency to varicose veins in the lower extremities.

In 1876 Moxon described two forms of latent albuminuria. One he called the *albuminuria of adolescence*. It occurs in youths and young men; the patients are languid, perhaps have headache; some slight derangement of the digestive system is often present; there is no evidence of organic disease, and if the urine were not examined the cases would be regarded as debility without any tangible signs of lesion. Albumen is usually found in the urine passed after breakfast, and is nearly surely present in some specimen of the urine collected for a period of forty-eight hours. A few hyaline casts and oxalates are often to be found. In the other form which he called *remittent albuminuria*, the albumen is present in greater quantity, usually after breakfast, but there is a remission at some period of the day, nearly always in the early morning, so that no albumen is present in the urine passed on rising after a night's rest. These cases have been lately re-named *cyclical albuminuria* by Dr. Pavy, but their relation to the erect posture is now admitted. It seems as if in the early part of the day the circulation in the kidney could not establish a proper equilibrium, but that after a certain number of hours this is attained, and no more albumen is lost until after rising the next morning.

Some years ago I published a series of cases of albuminuria in lads and young men, all suffering from dyspepsia.

Case 1.—T. S., æt. 15, labourer, complained of pain in the chest, and palpitation; had had a cold all winter; never had gonorrhœa; had measles six years ago; scarlatina twelve years ago; no other serious illness; he had never had dropsy. He was a tall, raw-boned lad, with a tendency to blueness about the cheeks and nose. His knuckles were bluish, and his hands and feet were often cold. Lungs normal, except slight prolongation of expiration. Heart not enlarged, apex in 5th interspace 1½ inches to left of sternum; first sound at apex reduplicated; second sound in aortic area accentuated; liver dulness 3½ inches in vertical mammillary line; appetite good; complained of feeling oppressed after eating; bowels confined; urine pale, clear, acid, sp. gr. 1009, a little albumen, no sugar or bile, one or two pale hyaline casts and a few epithelial scales (on two subsequent examinations no casts were found); ophthalmoscopic signs negative (Mr. Eales). The pulse was not diminished by 300 grammes pressure, but no satisfactory tracing was obtained. I saw this patient about a year after his case was published; he was looking much stronger, and his urine was free from albumen. I do not attach as much importance as I did to the signs of high arterial tension. These lads are nervous and excited under examination, so that the action of the heart is more forcible and frequent than normal, and, in consequence, the pulse is fuller and there may be some accentuation of the aortic 2nd sound; but I am convinced that their usual condition is one tending rather to low arterial tension.

It must be admitted that all our knowledge of these benign forms of albuminuria is recent, and in the face of authorities who maintain that even the smallest trace of albumen in the urine is always pathological, indicating, if not actual disease, a condition of renal stress which sooner or later tends to organic change, it is necessary to state what we know about their subsequent life history. Dr. Moxon mentions that all his cases ended in complete recovery. Most of mine have passed out of sight with persisting albuminuria, but some like the case just related have lost the symptom and appeared quite well. One of my medical friends at the age of 16 suffered from well marked remittent albuminuria; he lost it in four or five years, and has never had any trace of kidney disease, though it is now 21 years since the albuminuria was first noticed. I see him almost daily, and have often examined his urine. A gentleman was sent to me two years ago to be examined for life assurance; he was 46 years of age, and in every respect strong and sound except that his urine was albuminous; he told me that he was rejected for life assurance for this cause eighteen years ago, and that on two subsequent occasions his urine had been examined and found to be in the same state. There was no urethral discharge, and no other evidence of kidney disease. I may also refer to a young gentleman who was sent down from a public school in 1879 with albuminuria, and who was ordered by a London physician to winter in a warm climate as a case of Bright's disease, actually or potentially. This advice was not taken; he has continued to enjoy excellent health, and has always been a very active athlete. If he is going to have Bright's disease he has not yet shown any sign of it. Although I often see him I have not had an opportunity of examining his urine for some years; it formerly contained oxalates, but no casts.

What is the mechanism by which albumen passes into the urinary secretion? Since the time of Bright there have been two rival theories of this process, and of late years a third has been added; these three theories may be called: (1,) the Hæmatogenous; (2,) the Parenchymatous; and (3,) the Vascular. There always has been a school which attributed albuminuria not to changes in the kidney but to changes in the blood, and even in Bright's own day he was told that the structural alterations described by him were only the results of the elimination of albumen by the kidneys; this doctrine still survives and finds a persistent defender in Semmola, of Naples. His argument is especially directed to the etiology of Bright's disease, where he contends that the blood contains albuminoids of an abnormal diffusibility, so that they are found in the saliva, sweat and bile,—a fact in which he is supported by Vulpian and others. Tizzoni, however, found that albumen from the urine of a case of Bright's disease did not cause albuminuria when injected into the circulation of animals. Semmola entirely overlooks all that has been recently observed with regard to albuminuria apart from Bright's disease. Nothing can be farther from the truth than to suppose that the elimination of albumen by the kidney is liable to set up inflammatory action, for we know that it may go on for many years without any such result.

But variations in the diffusibility of the albuminoids of the blood may yet account for albuminuria in certain cases, and there is a great tendency on the part of authorities to accept such a view. Globulin diffuses more rapidly than serum albumen, and it has been stated by Lépine that albumen found in the urine after food diffused more rapidly than that passed fasting. The truth of the doctrine of hæmatogenous albuminuria is, by no means bound up with food albuminuria, so that it is perfectly allowable to doubt the occurrence of the latter, while admitting the general probability that changes in the blood account for a certain number of cases of albuminuria; for example in purpura, scurvy, profound anæmia, after hæmoglobinuria, etc.

With respect to food albuminuria we have Christison's case of the young man whose urine was always albuminous after eating cheese, and the experiments of Brown-Séquard, Barreswil and others in which the urine became albuminous on a diet largely composed of eggs. But in the first case the albumen excreted was not shown to be casein, nor in the second to be egg albumen, while more recent experiments have proved that it is not egg albumen* (GRAINGER STEWART), and that it is only when the digestive powers are overtaxed, as by swallowing many raw eggs together, that albuminuria occurs (LAUDER BRUNTON).

Parkes was one of the first to notice that there was an apparent increase of albumen after food. He stated that the albumen was increased after food in two cases of chronic Bright's disease, although it was diminished after food in a case of heart disease. He inclined to the view that the increase after food was due to the passage of imperfectly digested albumen, analogous to that which occurs when albumen is injected into the veins, or to the albumen undergoing some modification in the digestive process, such as its conversion into an acid albuminate, by which its diffusibility would be increased.

Parkes's facts do not seem to warrant this conclusion; for, while the albumen was increased after food in the two cases of Bright's disease, it was diminished under similar circumstances in the case of heart disease, so that it is necessary to postulate a peculiar inability to digest albumen in the former cases which did not exist in the latter. On the other hand, if we regard albuminuria as

---

* When egg albumen is injected into the veins, egg albumen is excreted by the kidneys.

a result of congestion of the kidney, we can easily understand why, in the cases of chronic Bright's disease, the stimulus to the circulation increased the albuminuria in already inflamed organs, while it diminished it in the case of heart disease, where the transudation was due to passive engorgement of the renal capillaries and veins. But Parkes's views have been adopted by many writers. Pavy supports Parkes by giving a table of the amounts of albumen excreted before and after breakfast in six experiments, all showing a marked increase. Pavy further adopted Parkes's view that this might be due to the increased diffusibility of the albumen, and supported it by showing that in some urines it is highly diffusible; but he did not attempt to prove that the albumen passed after food was more diffusible than that passed before food, or to indicate to what circumstances this increased diffusibility might be due.

Most physiologists deny that unchanged albumen is absorbed into the blood, and the only piece of direct evidence I know to the contrary is the statement of Brücke, that he has found coagulable albumen in the lacteals.

I endeavoured in the first place to determine whether this apparent increase of albumen after food really took place, and the following observations were made on a case of Bright's disease in the Queen's Hospital under the care of Dr. Carter. The patient had been on milk diet previous to this experiment, but on March 10th, 1880, he was put on the following diet:—

Breakfast, 5 a.m., two slices of bread and butter with tea.
Lunch, 10 a.m., two slices of bread and butter with one pint of milk.
Dinner, 1 p.m., four ounces of cooked meat, eight ounces of potatoes, bread and water.
Tea, 6 p.m., the same as breakfast.
Supper, 7 p.m., half-pint of milk.

He took walking exercise twice daily for half an hour at 12.20 and 2.30.

Date.	Period.	Quantity of Urine.	Albumen in 40 ccms.	Total Albumen.
March 10	7 P.M. to 5 A.M.	1620 cc.	·008 grms.	·1296 grm.
,,	5 A.M. to 1 P.M.	150 cc.	·016 ,,	·06 ,,
,,	1 P.M. to 5 P.M.	180 cc.	·01 ,,	·045 ,,
,,	5 P.M. to 7 P.M.	120 cc.	·008 ,,	·024 ,,
,, 12	7 P.M. to 5 A.M.	870 cc.	·0069 ,,	·1479 ,,
,,	5 A.M. to 1 P.M.	180 cc.	·017 ,,	·0756 ,,
,,	1 P.M. to 5 P.M.	180 cc.	·0141 ,,	·0630 ,,
,,	5 P.M. to 7 P.M.	150 cc.	·009 ,,	·0330 ,,
,, 13	7 P.M. to 5 A.M.	1410 cc.	·0155 ,,	·5358 ,,
,,	5 A.M. to 1 P.M.	180 cc.	·020 ,,	·090 ,,
,,	1 P.M. to 5 P.M.	180 cc.	·0166 ,,	·0738 ,,
,,	5 P.M. to 7 P.M.	120 cc.	·015 ,,	·0444 ,,

On each day, the relatively greatest quantity of albumen was excreted between breakfast and dinner. The quantity each day fell as the day advanced, in spite of the meat eaten at dinner and the exercise taken in the afternoon. The total quantity of albumen excreted rose steadily under the influence of meat diet, being three times greater on the fourth day than on the first.

One bottle containing the urine of part of March 11th got broken, so that day's record was incomplete. The analyses were performed by Dr. MacMunn.

This experiment is opposed to the doctrine of Parkes that unassimilated albumen is excreted by the kidneys. But it has been suggested that after food there may be a more rapidly diffusible albumen present in the blood. That there should be such differences is quite probable, as Graham long ago pointed out that an acid solution of albumen diffuses readily, while an alkaline solution scarcely diffuses at all; so that differences in the alkalinity of the blood serum may determine variations in the amount of albumen excreted. Pavy showed, using pericardium as a membrane, that egg albumen diffuses more readily than serum albumen, and Lépine has filled up the gap left by Pavy by observing that albumen in the urine after food diffused more rapidly than that passed fasting. The following experiments appear to show that the results may depend upon variations in acidity, etc., and not upon any primary difference in the albumen itself.

The septum used was vegetable parchment; and the time allowed was twenty-four hours in each case.

The following experiments were made on the urine of the same case:—

Experiment I.—*Before breakfast:* In bed. Urine faintly acid; one-third of a column of albumen; no albumen in diffusate.—*After breakfast:* In bed. Urine acid; four-fifths of a column of albumen; a trace of albumen in diffusate.

Experiment II.—*Before breakfast:* In bed. Urine neutral; a trace of albumen in diffusate.—*After breakfast:* In bed. Urine acid; a trace in diffusate.

Experiment III.—*Before breakfast:* In bed. Urine faintly acid; one-third of a column of albumen; a faint trace in diffusate.—*After breakfast:* In bed. Urine acid; two-thirds of a column; a trace in diffusate.

Experiment IV.—*Before breakfast:* In bed. Urine faintly acid; one-third of a column of albumen; a trace in diffusate.—*After breakfast:* In bed. Urine acid; two-thirds of a column of albumen; a trace in diffusate.

Experiment V.—*Before dinner:* Up. Urine neutral; a distinct cloud in diffusate.—*After dinner:* Up. Urine acid; a distinct cloud in diffusate, but less than before dinner.

Experiment VI.—*Before dinner:* Up. Urine acid; half a column of albumen; a cloud in diffusate.—*After dinner:* Up. Urine strongly acid; a whole column of albumen; a dense cloud in diffusate.

Experiment VII.—*Before dinner:* Up. Urine acid; half a column of albumen; a faint trace in diffusate.—*After dinner:* Up. Urine faintly acid; one-third of a column of albumen; a faint trace in diffusate.

As a rule, the albumen appeared to diffuse in proportion to the acidity of the urine. The diffusate bore no relation to the quantity of albumen present in the urine.

While it is fully admitted that more carefully conducted experiments might determine differences in diffusibility due as suggested to alterations in the salts of the blood or the alkalinity of the blood serum, there is no evidence that undigested albumen is ever excreted by the kidneys; and in this shape the doctrine of food albuminuria must be abandoned.

The parenchymatous theory ascribes the albuminuria to the destruction of the epithelial lining of the renal tubules, but there are several hypotheses which are strongly opposed to one another. One of the most interesting, that of Von Wittich, adopted by Ludwig and lately revived in Glasgow, is that albumen is physiologically transuded through the Malpighian tufts, but reabsorbed by the epithelium of the tubules. When the parenchyma is diseased this reabsorption is more or less hindered and albuminuria results. This theory, attractive as it is, is disposed of by the experiments of Posner and Ribbert, who proved by boiling freshly excised kidneys and hardening them in alcohol that there is no albumen present in the capsular space around the Malpighian tuft in healthy kidneys, though it can be easily demonstrated in albuminuric kidneys even when the epithelium is intact.

Another form of the parenchymatous hypothesis is that when the parenchyma has been shed from the tubules the basement membrane permits the transudation of albumen. This theory has never been disproved; and it is in its favour that kidneys prepared by the boiling and alcohol method showed albumen in the straight tubes ten minutes after ligature of the renal vein.

A third suggestion made by Senator is that the destruction of the renal epithelium itself furnishes a sensible amount of albumen. In the early stage of nephritis the epithelial cells shed their protoplasmic contents into the lumina of the tubules. This would only apply to cases of irritative or inflammatory albuminuria.

Lastly, there is the vascular theory. The ordinary seat of the transudation of albumen has been shown by Posner to be the Malpighian tufts; as was suggested by an experiment of Nussbaum's. In frogs the veins of the posterior extremities divide in the pelvis into two branches, one of which passes to the kidney like a portal vein, while the other joins its fellow of the opposite side to form the vena abdominalis anterior. The blood in the other branch passes through the kidneys and the liver into the vena cava inferior to reach the right side of the heart. The renal glomeruli receive their blood from the renal arteries, and the vasa efferentia pass into the same capillary network as that supplied by the renal portal veins, so that by tying the renal arteries in frogs the renal circulation is not brought to a stand-still as it is in mammalia. Taking advantage of this anatomical fact, Nussbaum tied the renal arteries, and then injected a five per cent. solution of egg albumen or a ten per cent. solution of peptone into the anterior abdominal vein without causing albuminuria, although when the renal arteries were not tied a smaller quantity sufficed to produce albuminuria. It is therefore proved that in frogs an albuminuric dyscrasia causes transudation of albumen only through the glomeruli, and it is probable that this is true also in man; taken together with Posner's observations we may regard the point as practically determined in that sense.

The question is what causes the transudation? Is

there a physical change in the membrane, or is there some alteration in the blood pressure or the rapidity of the blood current? These questions have given rise to much discussion. Thoma showed that in contracting kidney the walls of the glomeruli are actually abnormally permeable, permitting the passage, not only of thin fluids and colloids, but of small solids such as crystals of cinnabar, and this, too, in parts of the kidney presenting no recognisable structural alteration; but observations are wanting to enable us to extend this to other conditions under which albuminuria occurs. We know that very slight alterations in the coats of the vessels, such as may be induced by temporarily clamping or ligaturing the renal artery, give rise to albuminuria, and this may follow simple persistent blocking of the ureter. The action of certain poisons, for example, carbolic acid, has been proved, by boiling, and hardening the kidneys in alcohol, to cause albumen to transude through the glomerular wall without any visible structural alteration taking place (RIBBERT).

It must therefore be allowed that such changes may account for certain forms of albuminuria. But there are other important factors in this vascular theory, especially the pressure and rapidity of the circulation. Since Robinson experimented by tying the abdominal aorta below the origin of the renal arteries, the influence of increased blood pressure in determining albuminuria has been generally recognised; but of late years it has been called in question. It is fully admitted that the amount of urinary secretion varies directly with the pressure; but Runeberg has contended that the amount of albumen varies inversely. Lépine supports this in the following table :—

Pressure.	Albumen.
100 c.c.	1·3, 1·077, 0·66
40 c.c.	0·95, 1·5
100 c.c.	1, 0·7, 0·6
40 c.c.	0·8, 1·3, 1·4, 1·5

Senator accepts this view, and holds that the higher the pressure the more water but the less albumen. Bamberger also believes the albumen to be diminished as the pressure rises. Löbisch and Rokitansky caused albuminuria in healthy persons by lowering the blood pressure with pilocarpine. There can be no doubt that increased venous pressure leads to albuminuria as is seen clinically in cases of heart, lung, and liver disease, pregnancy, abdominal tumour, and probably too in many cases of debility, anæmia, and in the hypostatic congestion of fevers, pneumonia, &c. In ligature of the renal vein albuminuria occurs rapidly, and has been proved by Senator to take place by direct transudation into the tubules, probably from the lymphatics. Charcot and Bamberger attach much importance to the influence of slowing the current, which is shown by physical experiment to favour the filtration of albumen.

Ischæmia of the kidney, by experimental narrowing of the renal artery, leads to albuminuria, probably by slowing the current and reducing the pressure. Claude Bernard's puncture and experimental lesions of the spinal cord give rise to albuminuria by vaso-motor paralysis, the kidneys being deeply congested (SCHIFF). Many of the cases which have been regarded as neurotic albuminuria are susceptible of a more simple explanation, as has been already suggested for epilepsy and tetanus; and it will be admitted that other conditions are present in exophthalmic goitre, besides the nervous disturbance, sufficient to account for albuminuria, though this may be due in part to altered vascular innervation.

We may usefully summarise the teachings of this lecture in the following conclusions:—

1.—Albuminuria is defined as the presence in the urine of serum albumen, or serum globulin, or their modifications, syntonin and alkali albumen.

2.—Albuminuria may be present in healthy persons and

persist for long periods without causing any derangement of the general health, or of the structure of the kidneys.

3.—Albuminuria *per se* should not be regarded as an insuperable objection to life insurance.

4.—Albuminuria may occur in dyspeptic people, and in weakly over-grown persons, without being an indication of actual or potential renal disease.

5.—Albuminuria may depend upon many causes, grouped under three headings: (1), Hæmatogenous: due to alterations in the diffusibility of the blood albuminoids, owing to changes in the salts of the blood or the alkalinity of the blood serum; but albuminuria is never due, as has been asserted, to the excretion of undigested or partly digested albumen taken as food. (2), Parenchymatous: inflammatory changes in the epithelium give rise in the first instance to an albuminous exudate which must be present in the urine, and secondly, by destroying the cell layer and altering the basement membrane, allow direct transudation from the lymphatic vessels into the tubules. (3), Vascular: the walls of the glomeruli probably undergo alterations of their permeability from the effects of poisons, inflammation, and vaso-motor paralysis, while lowering of the blood pressure and slowing of the blood current favour filtration of albumen through them. In venous obstruction there is œdema of the whole organ and transudation of albuminous fluid direct from the lymphatic spaces into the tubules; in inflammation and vaso-motor paralysis a similar œdema is likely to occur with identical results.

BIBLIOGRAPHY.

BAMBERGER (VON). Ueber hämatogene Albuminurie. "Wiener Med. Woch.," 1881, Nos. 6 and 7.

BULL. Albuminuri—latent skrumpnyre ? " Nord. Med. Arkiv.," 1885, Bd. XVII., No. 25.

CHARCOT (J. M.). Conditions Pathogéniques de l'Albuminurie. " Le Progrès Méd.," Tome IX., 1881, p. 55.

Jones (Bence). Animal Chemistry in its Application to Stomach and Renal Diseases. London, 1850.

Lépine (R.). Quelques travaux relatifs à l'Albuminurie et à la Pathologie rénale. "Revue de Méd.," Tome II., 1882.

Leube (W.). Ueber die Ausscheidung von Eiweiss im Harn des gesunden Menschen. "Virchow's Archiv.," Bd. LXXII., Heft 2.

Marcacci (G.). Di un nuovo caso di Albuminuria fisiologica. "Comment. Clin. d. Mal. d. Org. Gen.-urin." Pisa. 1884, I.

Moxon (W.). On Chronic Intermittent Albuminuria. "Guy's Hosp. Reports," 3rd S., vol. XXIII.

Nussbaum (M.). Fortgesetzte Untersuchungen ueber die Secretion der Niere. "Pflüger's Archiv.," Bd. XVII., 1878, p. 580.

Pavy (F. W.). Gulstonian Lectures. On the Assimilation, and the influence of its defects on the Urine. "Lancet," 1862, II., p. 613; 1863, I., p. 461.

Posner (C.). Studien ueber pathologischen Exudatsbildungen. "Virchow's Archiv.," Bd. LXXIX., p. 311.

Ueber pathologische Albuminurie. "Berl. Klin. Woch.," Bd. XXII., 1885, p. 654.

Ribbert (H.). Ueber die Eiweissausscheidung durch die Nieren. "Cent. f. d. Med. Wiss.," 1879, p. 47 and 1881, p. 17.

Robinson (G.). Researches into the connection existing between an unnatural degree of compression of the blood contained in the renal vessels and the presence of certain abnormal matters in the urine. "Med. Chir. Trans.," 2nd S., Vol. VIII., 1843, p. 51.

Runeberg (J. W.). Ueber die pathologischen Bedingungen der Albuminurie. "D. Arch. für Klin. Med.," Bd. XXIII., p. 41.

Semmola. Nuove contribuzioni alla Patologia ed alla cura del Morbo di Bright. "Med. Contemp. Napoli," 1886, III., pp. 449 to 467. Also see "Le Progrès Médical," 1883, Tome XI., p. 471.

Senator (H.). Die Albuminurie im gesunden und kranken Zustande. Berlin: A. Hirschwald, 1882.

Stewart (Grainger). Clinical Lectures on important symptoms. Fasc. II. Albuminuria. Edinburgh: Bell and Bradfute, 1888.

Stirling (A. W.). Albuminuria in the apparently healthy. "Lancet," 1887, II., p. 1157.

Thoma (R.). Zur Kenntniss der Circulationsstörung in den Nieren bei chronischer interstitieller Nephritis. "Virchow's Archiv.," Bd. LXXI., Heft 1 and 2.

Thormählen (J.). Ueber eine eigenthümliche Eiweissart im menschlichen Urin. "Virchow's Archiv.," Bd. CVIII., p. 322.

Tizzoni (G.). Alcuni esperimenti intorno alla patogens dell' Albuminuria. "Gaz. deg. Ospit. Milano," 1885, VI. 12.

Wittich. Ueber Harnsecretion und Albuminurie. "Virchow's Archiv." Bd. X., p. 325.

Chapter II.

THE PATHOLOGY OF DROPSY.

Dropsy is certainly the most striking symptom of Bright's disease, and the one that commonly first attracts the patient's attention. Frerichs found it absent in only one out of 6·97 cases, Rosenstein in only one out of 80; but the proportion varies in the different forms of Bright's disease. It is usually present in acute nephritis, but it may be very slight or altogether absent. Thus it is commonly absent in cases of acute nephritis supervening in the course of pneumonia, diphtheria, typhoid fever, &c., though the state of the urine proves incontestably that acute nephritis exists. In subacute nephritis and in large white kidney, dropsy is very common, being absent in only eight or ten per cent. In the contracting form it is much less common; in the earlier stages it is generally absent, but later on it is more frequent. Thus, out of a hundred cases seen in the out-patient department, I found dropsy present in ten per cent. only, while in the bodies of persons who had died with contracting kidney, I found the proportion as high as twenty-five per cent. This variation is fully explained by the cause of the dropsy, which in contracting kidney must for the most part be attributed to heart failure. So long as the hypertrophied heart can do its work all goes on fairly well; but when this organ at last breaks down, dropsy appears along with other signs of the collapse of the whole system. Like ordinary cardiac dropsy, it is first seen in the legs and ankles, increasing after rising or getting about, and must be regarded as of grave prognosis, indicating failure of the heart to maintain the struggle any longer.

The fluid in renal dropsy varies in different situations, being poorer in albumen and solids in the subcutaneous tissue than in the serous sacs. The following table taken from Bartels gives the composition in the various seats of dropsy.

	Sp. g.	Water	Solids	Inorganic Salts	Albumen
Blood Serum	1015·58	957·656	41·402		30·39
Pericardial fluid	1009·69	978·622	21·38	15·58	?
Peritoneal ,,	1009·63	984·31	15·69	14·48	2·55
Subcutaneous ,,	1007·65	988·30	11·70	11.69	

Schmidt gives the following figures:—
Subcutaneous tissue, 0·36 % albumen; Meninges, 0·6 to 0·8 %; Peritoneum, 1·13 %; and Pleura, 2·85 %.

Urea has been found in it by many observers, and Bartels quotes the following figures from Edlefsen:—
Subcutaneous tissue, 0·359 %; Peritoneum, 0·28 to 0·30 %; Pericardium, 1 %.

What is dropsy? Dropsy consists in an accumulation of watery fluid in the lymph spaces of the subcutaneous cellular tissue and in the serous cavities of the body. This fluid or lymph is derived from the capillaries, and under normal circumstances is poured out into these spaces, but taken up again by the venous and lymphatic radicles as fast as it is poured out.

For the production of dropsy, the equilibrium of this arrangement must be upset by either (*a*) an increase in the out-flow of fluid, or (*b*) a failure on the part of the veins and lymphatics to take up the effused fluid.

An increase in the out-flow of lymph from the capillaries is caused experimentally by section of the vasomotor nerves, and occurs in disease when their function is paralysed; it may also be due theoretically to increased permeability of the vascular walls, and to changes in the blood serum. We shall afterwards see what share to assign to each of these factors.

The principal source of power in pumping these lymph spaces dry is the heart; it is assisted by the aspiratory action of the thorax in respiration, and the contractions of the muscles in the limbs. When the heart fails to maintain a negative pressure in the veins œdema sets in. Localised venous obstruction acts with less certainty in the same way, as the lymphatics which open into the venous circulation beyond the obstruction may carry on the work; if, however, their task is added to by cutting the vaso-motor nerves, and so leading to a greater influx of blood to the capillaries and increase in the outflow of lymph, dropsy sets in. Section of the nerves probably has more than this simple effect, but it need not concern us at present. There are no difficulties in the way of understanding dropsy due to heart failure; it is a break down in the central pumping apparatus, and need not detain us longer.

The dropsy of acute nephritis presents problems of greater difficulty. In most cases there can be no question of heart failure, and its explanation must be sought for in the factors that determine an excessive outpouring of lymph from the blood vessels. In acute Bright's disease the urinary secretion is very scanty, and as patients continue to swallow fluids the total quantity of blood must increase absolutely in volume while suffering a relative diminution of its solids; that is to say, the water is absolutely increased and the solids relatively decreased, inducing a state of hydræmic plethora.

A watery condition of the blood has long been regarded as a cause of dropsy, some authors seeing in it a predisposing factor, others ascribing to it still greater importance. Brücke's experiment of cutting one sciatic nerve of a frog and putting the animal on a piece of moist blotting paper in a large glass vessel, showed that after some weeks œdema occurred in the palsied limb, which disappeared on food being administered, reappeared in inanition, and disappeared again when nutrition was restored. But there is considerable difficulty in ascertaining what are the circumstances under which hydræmia will produce dropsy, for authorities certainly differ as they are wont to do.

One of our earliest experimenters, Hales, succeeded in producing "both the ascites and the anasarca" in a dog by pouring water into the jugular vein, but when we read the description it appears that the dog died during the progress of the experiment, and we are not told whether these phenomena preceded death or not; this is an important point, as the retentive power of the vascular walls alters altogether after death. Magendie in his "Leçons sur les Phénomènes Physiques de la Vie" alludes to this matter in several lectures; he certainly tells us that he succeeded in producing general dropsy by defibrinating the blood of a dog, and in another experiment he produced analogous results by pouring water into the veins, as Hales had done. But he performed a much more remarkable experiment by introducing no less than ten litres of water into the veins of a patient suffering from hydrophobia; he had observed the wonderful calmative influence which similar injections had upon ferocious dogs, and he hoped by this means to combat this fatal disease. His patient lived several days, and died presenting the phenomena of several "pseudarthroses" or dropsical swellings of joints. We are not told that œdema was absent, but from the relation the story bears to the main point of the lecture, we may feel justified in concluding that had there been any anasarca, the lecturer would not have failed to mention it.

Niemeyer states—I know not on whose authority—that "if we abstract blood from an animal and inject a corresponding quantity of water into its veins in its stead, the animal does not become dropsical"; on the other hand, Jaccoud says "the artificial dyscrasia created by the injection of water is rapidly and constantly followed by temporary dropsy." Wagner quotes Donders, Kierulf and Hermann in proof that dropsy may be caused by the injection of abundance of water into the circulation; but to this is opposed the experiment of Cohnheim, which showed that a solution of salt might be passed under low pressure through a limited vascular area, such as the ear of a rabbit, without causing œdema. It is obvious that all these experiments do not fulfil the conditions necessary to throw light upon the influence which hydræmia may have in producing dropsy in certain diseases. Injections of pure water destroy the blood corpuscles, and paralyse the heart, as well as, in all probability, the walls of the blood-vessels; while the pressure employed by some of the observers was probably greater than the normal blood pressure of the animal.

Bartels ranged himself very unreservedly on the side of those who regard the hydræmia of Bright's disease as the efficient cause of the dropsy, but he thought that there had been much confusion in the minds of many physicians who supported this view, as to the cause of the watery state of the blood. He admitted that in renal diseases a large quantity of serum albumen is lost by the blood, and he believed that this must render the blood serum more watery; still, he was of opinion that this is neither the sole nor the essential cause of the hydræmia, for renal disease may lead to most extensive dropsy before any quantity of albumen worth mentioning has been eliminated; and he regarded the occurrence of dropsy as dependent less upon the loss of albumen than upon the diminution of the amount of water secreted by the kidney. Even in cardiac disease, he said, we see the influence of the quantity of the renal secretion on the presence of œdema, and he denied that the malnutrition existing has anything considerable to do with the matter. On the other hand, in renal disease a considerable daily loss of albumen may take place without any dropsy resulting if only the quantity of water passed is sufficiently copious, while dropsy may ensue during an insignificant loss of albumen directly the daily urinary secretion falls below a certain minimum quantity.

He quoted the observations of Rehder, who, with the view of establishing the proportion borne by the quantity of water taken in drink and food towards that which afterwards appears in the urine, instituted a series of experiments upon healthy persons living under identical external conditions and then compared the results with a series of analogous observations made on dropsical subjects.

He subjected five healthy persons to experiment, and the average of the daily results, extending over thirty-two days, was, that of every hundred parts of fluid ingested, 76·4 parts were excreted in the urine; the lowest mean in one individual was 68 per cent., the highest 88 per cent. Very contrary results were obtained by the observations on persons suffering from cardiac and renal dropsy. A man aged fifty, with cardiac disease, excreted quantities of water varying from 29·7 to 49·2 per cent.; a youth with chronic parenchymatous nephritis excreted 18·6 to 33·5 per cent. only; in a woman with the same disease the excreted water averaged 16 per cent., and never exceeded 24·4 per cent.; a man with the same gave an average of 28 per cent. In another case the exact correspondence between the increase and subsidence of the anasarca and the diminution and increase of the urinary secretion was perfectly established.

Bartels held that these experiments show that the dropsy of renal disease is due to the relative insufficiency of the kidneys to eliminate water; but some time must elapse before anasarca occurs; in one case, in which stoppage of both ureters was present, one hundred and twenty-two hours elapsed without a trace of œdema showing itself, but this patient vomited large quantities of liquid.

The cases given by Trousseau of two patients suffering from general anasarca due simply to retention of urine, and whose symptoms were completely relieved by emptying the bladder, are, if any be needed, additional clinical illustrations of the dependence of dropsy on a free exit of water by the kidneys.

Cohnheim and Lichtheim directly oppose Bartels' view, and they have endeavoured to show that hydræmia *per se* cannot produce subcutaneous œdema. Their experiments were made by injecting, with a syringe or under very low constant pressure, large quantities of a blood-warm 0·6 per cent. solution of common salt into the circulation of dogs, rabbits, and other animals; rabbits were used without any preparation, but dogs and the larger animals were soothed with morphia or curarised. As a preliminary proceeding they endeavoured to gain some idea of the quantity of fluid which might be injected without killing the animal, and they found that dogs could stand much more than rabbits; the latter generally could support an injection of 46 per cent. of their body weight before they died, but dogs withstood a much greater quantity, especially if the abdominal cavity were opened, 92 per cent. of his body weight being introduced into one dog before he died. In one case the dog died of acute œdema of the lung, but they were unable to account for this exceptional occurrence; in general the animals died with symptoms of deoxygenation of the blood, paralysis of the heart, and in some cases convulsions.

In none of these experiments—and they were very numerous—was any subcutaneous œdema or anasarca produced; even after the largest injections the subcutaneous tissue was quite free from water, and as this is the earliest and principal seat of the so-called hydræmic œdema, it appeared from these observations that hydræmic plethora had nothing to do with hydræmic œdema. The following figures show the amount of hydræmia they produced:— They found the normal dried residue of a dog's blood to be about 20 per cent., the highest being 24·09 per cent.; Andral and Gavarret found the dried residue of the blood of a case of kidney disease to be 13·23 per cent.; in their experiments the percentage fell from 22·05 per cent. to 11·64 in one case; in another from 20·76 to 11·33; in a third, with the abdomen opened, from 22·16 to 8·29 per cent. A still greater degree of hydræmia was produced by ligaturing the whole portal circulation; for example, in one case in which the cœliac, superior and inferior mesenteric arteries and the portal vein were ligatured, the percentage fell from 24·09 to 4·87.

Many authors have considered the hydræmia of renal dropsy as it affects not the entire solids of the blood but those of the blood serum. Christison found these to be 6·1 per cent., and Frerichs, in three cases of commencing nephritis, found them 9·19 per cent. In one of their cases, after injecting 64 per cent. of the body weight of common salt solution the serum residue was only 2 per cent., so that there can be no doubt that the thinning of the blood produced in these experiments not only equalled but exceeded that which occurs in disease, and the absence of anasarca cannot be thus explained.

In the next place they proceeded to enquire what influence hydræmic plethora had upon the pressure and rapidity of the blood current; this is of especial interest with reference to the views of Traube, who regarded the increased arterial tension and the hypertrophy of the left ventricle as the result of the diminished secretion of water, and the resulting hydræmic plethora. They could not, succeed in raising the pressure above the normal except momentarily, and they found important increase of pressure to occur only when it had previously been below normal. The shape of the curve of the blood-pressure underwent remarkable changes during the experiments; at first the respiratory wave disappeared and only the pulse wave remained, but got much larger; after a time the curve resumed its original outline. A similar negative result followed experiments on the venous pressure; a slight increase was produced temporarily, and was accompanied by a venous pulse proved to originate by back-stroke from the heart. The blood current was *quickened*; this was directly observed in the mesentery and tongue of frogs, but it did not last long. They observed the same in the mesentery of dogs and rabbits, and also by counting the blood drops from the vein of the anterior extremity of a dog; although this method cannot be exact, the increase was too enormous to leave any doubt.

The most obvious appearance was transudation of water from the blood, not only through the glandular organs, but into their tissues themselves. The first point seems to indicate that all glandular organs secrete a very large quantity of water. The animals passed a very large quantity of pale limpid urine, which in some rare cases contained a small quantity of albumen, and more frequently but not constantly, sugar, which had been previously recorded by Bock and Hoffmann. The saliva and the secretions of the glands of the mouth were increased enormously, also the secretions of the gastric and intestinal glands; the bile was richly secreted, but the pancreatic fluid was not increased. The secretion of the conjunctiva, lachrymal glands and mucous membrane of the nose was greater than normal, and the perspiration in two experiments on a goat and on a horse was very profuse.

In order to estimate the part played by the absorbents, a cannula was placed in the thoracic duct and an enormous acceleration of the lymph stream was demonstrated, but it is remarkable that when the cannula was placed in the lymphatic trunk of an extremity no acceleration of the lymph stream could be produced; between these two extremes the lymphatics of the neck showed a definite increase, but not nearly equal to what was observed in the thoracic duct. These differences correspond to the differences produced by hydræmic plethora in different parts; for although no œdema of the subcutaneous connective tissue was ever observed, there was a special localisation of the dropsy which the authors believed to be characteristic of the dropsy of hydræmic plethora.

The bodies of the animals swelled very much, and the peritoneal cavity was always full of fluid, as were the intestines, while their mucous and submucous coats were œdematous; the coats of the stomach presented a still higher grade of œdematous swelling. The lymphatics and chyle ducts of the mesentery were distended and the mesenteric lymph glands were swollen and œdematous. The pancreas presented the most extreme degree of dropsical distension; the kidneys were both enlarged, pale, and extremely moist on section; the liver was distended, of doughy consistence, and on section discharged a large quantity of thin watery blood; the gall bladder was very œdematous, its wall being more than a millimetre thick. The vesiculæ seminales and the retroperitoneal tissue were also œdematous. The spleen was somewhat swollen and tight to the feel, but frequently scarcely at all enlarged. The thoracic cavities and viscera presented a striking contrast; neither pericardium nor pleural cavities contained a drop of fluid; the lungs (except in one case, in which they were œdematous) were quite dry, or at most there was only slight œdema of the connective tissue bands radiating from the hilus. The sub-maxillary, salivary and lymphatic glands were swollen, as were the sublingual glands and the lymphatic glands of the neck. The conjunctivæ and lachrymal glands were œdematous. All other organs, such as those of the central nervous system, were free from œdema.

These results were uniformly the same in all their numerous experiments. In order to exclude the possible objection that by injection into the jugular vein, the fluid was driven directly into the inferior cava and the veins of the liver so as to cause a special obstruction of the portal circulation, they injected fluid into the femoral vein or into an artery, and got the same results.

They varied their experiments by using other fluids; distilled water killed the animals too soon by destroying the blood corpuscles, but solutions of sugar, various salts, albumen and blood serum gave the same results as the solution of common salt. They also used defibrinated dog's blood in a few cases, and, except that the resulting œdema was much less, often somewhat sanguineous, and complicated by punctiform hæmorrhages, the results were the same as before.

Again, they bled dogs and injected an equal quantity of salt solution, without causing anything at all except slight œdema around the operation wound. But, as it might be questioned whether the skin of dogs and rabbits is analogous to that of man, more especially as the important sweat function of the latter is absent in these animals, they performed experiments on a horse and a goat. Both animals died of œdema of the lung; before death there was copious perspiration, but no subcutaneous œdema.

The following circumstance appeared to throw some light upon the occurrence of œdema. In fixing their dogs for experiment they used to place an iron ring between their toes and fasten this to their noses by a loop; they noticed that the snouts of the animals became œdematous; they repeated the experiments on animals with wounds, and found that œdema occurred around the wound. Painting with iodine always produced slight subcutaneous œdema, but animals so treated showed marked œdema after hydræmic plethora was induced. They produced the most superficial dermatitis by shaving animals and exposing them to the sun's rays for an hour or two, and the slight œdema resulting was greatly increased by hydræmic plethora. Tying the femoral vein of a dog produces no œdema of the corresponding foot, but this occurred frequently when the blood was thinned.

They argue from these facts that some change in the state of the vessels is the necessary factor for the occurence of œdema in hydræmia; and while they admit that it is not easy to explain renal dropsy by any demonstrated changes in the cutaneous vessels, the combination in scarlatina is very noteworthy. In this disease along with the skin affection there is often slight œdema, which becomes very pronounced on the supervention of suppression of urinary secretion by renal disease and the consequent hydræmic plethora; occasionally the changes in the skin vessels are sufficiently great to cause œdema, apart from renal mischief, forming the long recognised *hydrops scarlatinosus* without albuminuria.

The frequent absence of dropsy in the acute nephritis of pneumonia, diphtheria and other non-eruptive febrile diseases already alluded to, supports this explanation.

In the hydræmic dropsies of cachectic diseases they suppose that the prolonged hydræmia injures the walls of the vessels.

Although it is competent to deny the parallelism of these experimental conditions with those of disease, we must admit that they corroborate the previously alluded to experiment of Cohnheim, that watery solutions of common salt at the temperature of the body do not readily pass through the walls of the subcutaneous vessels; and there seems reason to doubt whether, volume and pressure remaining the same, simple hydræmia can by itself induce œdema. We know that inanition rarely produces dropsy in the human subject; patients suffering from malignant stricture of the œsophagus often present the most extreme degree of emaciation without any or with the slightest possible dropsical swelling. Virchow noticed no dropsy in the famine in the provinces of Upper Silesia and Spessart; individuals who suffer from most profound anæmia from hæmorrhages rarely show corresponding œdema.

Many cases of total suppression of urine from calculous obstruction have been observed without œdema, but there is frequently constant vomiting which would tend to prevent an increase in the volume of fluid in the circulation.

In hysterical oliguria there may be an extreme reduction of the urinary secretion without œdema or vomiting. In a case seen by me with Mr. Lawson Tait in September, 1885, a lady passed only 16 ounces of urine in a week, without dropsy, vomiting, or uræmic symptoms. She was under close observation in Mr. Tait's private hospital, so that there can be no room for doubt as to the facts.

We are therefore driven to look to a third factor, increased permeability of the vessels, to fully explain the dropsy of acute nephritis. What the exact nature of the physical change is, remains obscure. Cohnheim, perceiving its analogy with what occurs in inflammation, calls it an "inflammatory" change; he believes that it may be brought about when hydræmia has persisted for a long time. It is probably influenced by the innervation of the blood vessels. Landois says it occurs when the blood contains dissolved hæmoglobin or too little oxygen or albumen. Lauder Brunton has found that the permeability of the vessels after death is increased by the presence of acids in the blood, and suggests that acids or substances acting in the same way may accumulate in the blood in Bright's disease. He quotes from an inauguration thesis by Feitelberg, who shows that a number of poisons, among them arsenic, have the power of increasing the acidity of the blood; and he recalls the œdematous condition of the eyelids induced by arsenic as an illustration to the point. This is all that is at present known on the subject.

Summary.—1. Dropsy is an accumulation of lymph or watery fluid in the lymph spaces of the body as a consequence of—(*a*) defective pumping arrangements, by which the fluid is allowed to accumulate; (*b*) changes in the blood and capillaries by which the outflow of fluid is increased.

2. Dropsy is generally present in acute and subacute nephritis, and in about 90 per cent. of cases of large white kidney; but in uncomplicated contracting kidney it is absent in at least 90 per cent. of those able to get about. In the later stages of contracting kidney it is more frequent, being present in 25 per cent.

3. In uncomplicated contracting kidney dropsy is due to heart failure; but in acute nephritis and its sequelæ it is due to more obscure causes in which a watery state of the blood is accompanied by increased permeability of the capillaries, caused possibly by the presence of an acid or other toxic substance in the blood.

BIBLIOGRAPHY.

BARTELS (C.). Art. Dropsy. "Ziemssen's Cyclopædia of the Practice of Medicine," Vol. XV.

BRUNTON (LAUDER). Pathology of Dropsy. "Practitioner," Vol. XXXI., 1883, p. 177.

COHNHEIM (J.). Embolische Prozesse.

COHNHEIM AND LICHTHEIM. Ueber Hydrämie und hydrämisches Oedem. "Virchow's Archiv.," Bd. LXIX., 1877, p. 106.

HALES (S.). Statical Essays. Vol. II., p. 110.

JACCOUD (S.). Pathologie interne. Tome I., p. 53.

LANDOIS (L.). A Text-book of Human Physiology. Stirling's translation. Vol. I., p. 420.

NIEMEYER (F. VON). Text-book of Practical Medicine. Vol. II., p. 29.

WAGNER (E.). Manual of General Pathology. Amer. Edition, 1876, pp. 234 and 542.

Chapter III.
PATHOLOGICAL RELATIONS OF TUBE CASTS.

Since it has been shown beyond dispute that albuminuria may be present not only in acute diseases, but in various chronic maladies, apart from renal inflammation, and even under certain circumstances in healthy men, it has lost its former significance as evidence of Bright's disease.

Moreover, we know that Bright's disease may be present, and progress to its termination (Bartels) without the occurrence of albuminuria; so that we stand greatly in need of some more trustworthy sign.

If in recent years the discovery of tube casts in the urine of patients suffering from jaundice, diabetes and secondary congestions of the kidneys, or even in more transitory conditions, as in the urine of the famous pedestrian Weston, during his prolonged walk, may have appeared to throw a little doubt on their diagnostic value, it is certain that they afford us the best and clearest indications of the changes that are taking place in the renal epithelium,—changes which rightly interpreted afford the safest grounds for diagnosis and prognosis.

Varieties of Casts. Casts of the renal tubules are of three kinds: (1,) Blood Casts; (2,) Epithelial Casts; (3,) Hyaline Casts. (*See* page 38.)

There are besides casts compounded of these varieties, blood or epithelium adhering to or forming part of a hyaline cast; while the term "granular" is frequently used to describe either epithelial or hyaline casts, which have become opaque and granular from infiltration with fat granules or micro-organisms.

Blood Casts. Casts composed of blood clot, *i.e.*, blood corpuscles matted together with fibrin (*Fig.* 1, *a*) indicate as may be supposed, the entrance of blood into the tubules of the kidney. The source of the hæmorrhage is for the most part the capillary tufts of the Malpighian bodies. The blood in its passage down the tubes becomes coagulated, and passes into the urine in the shape of cylindrical casts of the tubules. The presence of such casts affords important evidence in hæmaturia, proving that the blood comes from the glandular substance of the kidney, and

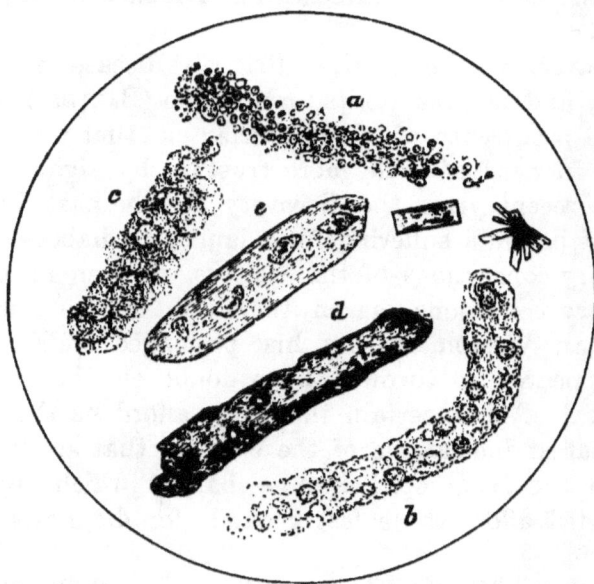

Fig. 1. Varieties of tube casts: *a*, Blood cast; *b*, Epithelial cast composed of small round cells; *c*, Epithelial cast formed of desquamated and fatty epithelium; *d*, Granular hyaline cast; *e*, Hyalo-epithelial cast,

not from the renal pelvis as in calculus of the kidney, or from any lower part of the urinary tract. These casts are seen in the early stages of acute nephritis, so long as hæmaturia persists, or when it occurs in any form of Bright's disease, as well as in hæmorrhage from congestion. They must therefore be regarded as evidences only of rupture of blood vessels in the glandular part of the kidney, and not as signs of any stage or form of Bright's disease.

Epithelial Casts. There are two main types under which the renal epithelium appears in the urine under the form of casts of the tubules of the kidney. In one the cast is made up of distinct small round granular cells, like leucocytes, but of smaller size, and which are, as their appearance suggests, proliferated renal epithelium (*Fig.* 1, *b*).

These are met with in acute and sub-acute nephritis, and always indicate an active degree of inflammation, with proliferation of the renal epithelium.

In the second type the cast is composed of a mass of epithelial cells, crowded together so as to obscure their individual outlines, and more or less opaque from fatty degeneration (*Fig.* 1, *c*).

These are formed by the desquamation of the epithelium, which is pushed off the basement membrane by the inflammatory exudation from the venous plexuses surrounding the tubule; they are usually of large diameter, being moulded in tubes denuded of epithelium, and they witness to severe diffuse inflammation of the kidney. They are met with in cases of recent inflammation in healthy kidneys, in sub-acute nephritis, and in the acute and sub-acute attacks which so frequently supervene in the course of the contracting kidney.

Hyaline or Colloid Casts. These are by far the most common casts; they are met with in all forms of Bright's disease, in venous congestion, and in many conditions where the kidney is undergoing slight or temporary irritation. They occur in jaundice, diabetes, heart disease, and pregnancy; they were found in the urine of Weston during his famous walk, and are occasionally seen in the urine of dyspeptic patients, especially when oxaluria or lithuria is present.

They are usually slender homogeneous transparent cylinders (*Fig.* 2, *a*), but are often more or less opaque from fatty degeneration, and are then called "granular" casts (*Fig.* 1, *d*).

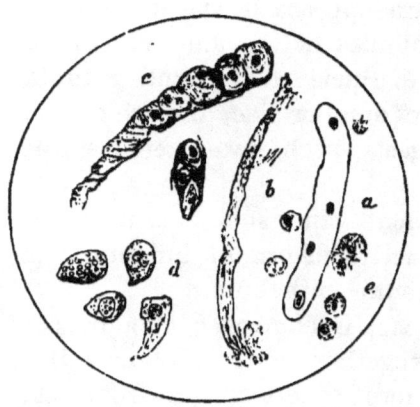

Fig. 2. Deposit from acute catarrhal nephritis showing: *a*, Slender hyaline cast; *b*, Mucous cylinder; *c*, Hyaline and Epithelial cast; *d*, Pear-shaped Epithelial cells from pelvis of kidney; *e*, Epithelium from tubules.

Various origins are assigned to them, and they are probably formed in three different ways.

Langhans has described a colloid metamorphosis of epithelial casts by which they become wholly or partly hyaline or glass-like in appearance; if the change is only partial they would be called hyalo-epithelial casts (*Fig.* 1, *e*); if complete they form one kind of hyaline cast, which, like the epithelial cast it originally was, is distinguished by its large diameter. In some instances, these transformed casts give the characteristic colour reactions of lardaceous material, turning mahogany-brown with iodine and rose-violet with methyl blue; but not necessarily in association with or dependent upon lardaceous changes in the blood-vessels of the kidney.

The second mode of origin is by coagulation of transuded plasma. Salkowski and Leube suggest that fibrin is formed by the action of the dead epithelium on the fibrinogen of the blood serum. This doctrine has been warmly opposed by the French school who have pointed out that hyaline casts are physically and chemically differentiated from fibrin; they are non-fibrillated, they do not swell upon the addition of acetic acid, and they are soluble in distilled water, especially when warmed. Fibrin cylinders are sometimes met with, but supporters of the doctrine that hyaline casts, or some variety of them are derived from the blood, ascribe them now to a modification of albumen, effected in some such way as acidulation by the renal epithelium (RIBBERT).

This experimenter tied the renal artery of a rabbit for $1\frac{1}{2}$ hours and then injected a weak $2\frac{1}{2}$ per cent. solution of acetic acid into the jugular vein. After boiling the excised kidney he found Bowman's capsules filled with a hyaline mass, and after hardening in alcohol this mass was granular. He also boiled an albuminuric kidney in acidulated water, with the result that beautiful hyaline masses were formed in it; uric and phosphoric acid and hydrochloric acid gave equally good results; but urea did as well, so that acidulation *per se* appears not to be the essential condition. As moreover, these casts have been found experimentally in kidneys whose epithelium was quite intact, it must be admitted that congestion alone may give rise to them.

A third theory ascribes them to an exudation or secretion from the tubular epithelium. Aufrecht found after ligature of the ureter in rabbits, if the kidneys were examined within the first three days, the tubules contained many hyaline cylinders, although the epithelium was intact and the interstitial tissue and blood vessels did not show the slightest change; moreover, he once saw a cylinder made up of single irregular pieces, separated by fine bright lines, and some of the epithelial cells had fine bright rounded structures projecting from them.

Strauss and Germont have recently confirmed this observation by researches made with a degree of care that leaves nothing to be desired.

In thin sections of pieces of kidney hardened in osmic acid, it is seen that even in the first few hours after the ligature the clear protoplasm of the epithelium of the convoluted tubes, and of Henle's loops disappears, and only the striated basal portions with the nuclei remain.

This change is closely connected with the formation of these casts, which three or four days after the ligature are present in great numbers in the convoluted tubules, as well as in the medulla and pyramids. Other tubules may be seen not yet filled by casts, in which the granular contents of the epithelial cells are projecting into the lumen, in the shape of spherical processes of the same colour and consistence as the casts. The following stages in the formation of casts from the exuded droplets can be noted: some are already exuded from the cells and lie free in the lumen of the tube (*Fig.* 3); others project half-way or more and remain still in connection with the cell matrix.

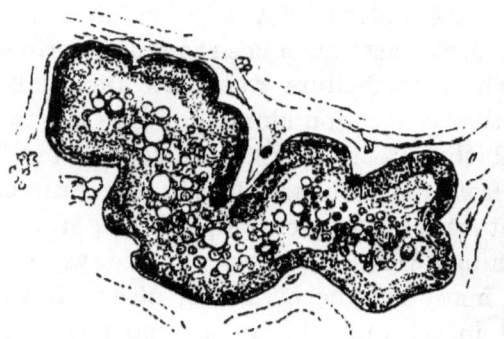

Fig. 3. Convoluted tubule, catarrhal nephritis. Epithelium represented by only basal portion. Cavity of tubule filled with protoplasmic droplets from which casts are formed.

Where the change is further advanced, a number of droplets may be seen to group themselves so as to form a cylindrical mass, the centre of which is homogeneous while the exterior is scalloped by the want of coalescence of its spherical components.

It is said by Kelsch that casts formed in this way are composed of a colloid material which differs from that of the ordinary hyaline cast, by coagulating into a harder and more brittle substance, having a feebly yellow colour, which stains dark brown with osmic acid, and rose-pink with picro-carmine.

There appears to be good reason for accepting the view that hyaline casts originate in each of these ways. It has been already stated that those formed by metamorphosis from epithelial casts, can generally be recognised by their larger size, or by the change being incomplete. Of the other two kinds it is a question which is the more common. On *à priori* grounds we should expect that where there is only congestion the casts should be formed by transudation; but where inflammatory conditions are present, by secretion. If this is really so, the staining reactions given by Kelsch would afford valuable indications.

I believe that casts are very rarely formed by transudation, and my opinion is based on their rarity in so-called functional albuminuria. As a clinical fact we know that casts are only few and far between, unless inflammation is present. We know that in diabetes, where casts are present, there is apt to be structural kidney change; and their presence in jaundice is explained by the observations of Mobius, who found that persistent excretion of bile by the kidneys destroyed the renal epithelium. So, too, in old standing heart disease, the kidneys are liable to more or less change of a chronic inflammatory nature. It is also quite probable that there may be transient conditions of renal irritation, comparable with that produced by temporary constriction of the renal artery, in which the epithelium may secrete casts, as in the case of Weston, and in attacks of gout. Practically, therefore, we ought to attach very high importance to the recognition of epithelial and hyaline casts as evidences of irritation or inflammation of the renal epithelium. We can measure the intensity of this process by their number, and watch its progress by their persistence.

Epithelial casts bear witness to severe inflammation, such as occurs in acute and sub-acute Bright's disease; broad hyaline casts have the same meaning, as they are formed from the epithelial variety as already explained, but slender hyaline casts testify only to a mild inflammatory process, such as occurs in the contracting kidney, or during the subsidence of an acute attack.

Summary.—1. Casts of the renal tubules are of three kinds—blood, epithelial, and hyaline.

2. Blood casts are evidence of hæmorrhage from the glandular substance of the kidney.

3. Epithelial and broad hyaline casts are evidence of acute and sub-acute inflammation of the kidney.

4. Slender hyaline casts are evidence only of a mild inflammatory process, which may be the result of a transient irritation.

BIBLIOGRAPHY.

AUFRECHT. Die diffuse Nephritis und die Entzundung in allgemeine. "Cent. f. d. Med. Wiss.," 1878, p. 337.

CORNIL and BRAULT. Etudes sur la Pathologie du Rein. Paris: Félix Alcan, 1884.

LANGHANS (T.). Ueber die Veränderungen der Glomeruli bei der Nephritis nebst einigen Bemerkungen über die Entstehung der Fibrincylinder. "Virchow's Archiv.," Bd. LXXVI., p. 85.

MOBIUS. Beiträge der Niere beim Icterus. "Arch. d. Heilkunde," Bd. XVIII., p. 83.

POSNER (C.). Studien über pathologischen Exudatsbildungen. "Virchow's Archiv.," Bd. LXXIX., p. 311.

RIBBERT (H.). Ueber die Eiweissausscheidung durch die Nieren. "Cent. f. d. Med. Wiss.," 1879, p. 47, and 1881, p. 17.

STRAUSS and GERMONT. Des Lésions histologiques du Rein, chez le Cobaye, à la suite de la Ligature de l'Uretère. "Arch. de Phys.," Tome IX., 1882, p. 386.

Chapter IV.
CARDIO-VASCULAR CHANGES.

Next to dropsy and the state of the urine, the alterations in the circulatory system are the most striking clinical and pathological facts met with in Bright's disease. Clinically, they are present in the shape of physical evidences of cardiac hypertrophy, with increased arterial tension; pathologically, the changes described are hypertrophy of the heart, and thickening of the small arteries.

The hypertrophy of the heart may be a genuine hyperplasia of the muscular substance, but evidences of interstitial myocarditis, such as thickening of the intermuscular fibrous tissue, and pigmentary or fatty degeneration of the muscular fibre are often present.

The heart may enlarge very rapidly even in acute nephritis, but it is in association with the chronic forms, especially the contracting kidney, that hypertrophy is most commonly met with. Bartels asserted that he had "never failed to obtain the objective signs of hypertrophy of the left ventricle in any of his cases of genuine contracting kidney, or to confirm the fact in every *post mortem* made upon their bodies, and therefore he can never hold that a diagnosis of this renal disease is made certain when no enlargement of the left ventricle is recognised." This statement would be of the utmost value if true, for we should then have a most important confirmatory sign.

Dickinson says that an analysis of 250 cases of granular degeneration, drawn from St. George's Hospital books, gave 48 per cent. as the proportion of cardiac enlargement, and that since his attention was directed to the subject he has scarcely seen an instance in which, if the renal disease was distinctly recognised, whether after death or in life, some degree of cardiac hypertrophy was not also present. He regards simple cardiac hypertrophy as one of the most important diagnostic signs of this form of renal disease.

Ewald found hypertrophy of the heart in 20 out of 21 cases of granular kidney. On the other hand Grainger Stewart speaks of nearly one half of a series of cases he examined *post mortem* having had enlarged heart simply from renal disease, while many others had enlargement connected with valvular or vascular lesions.

Gull and Sutton state that they have particulars of nine cases in which the kidneys were very contracted and the heart was free from hypertrophy.

I found that out of 100 typical cases examined *post mortem* at the General Hospital 44 had simple cardiac hypertrophy, and 16 hypertrophy with valvular disease, the proportion being 60 per cent. of hypertrophy with and without valvular disease. In addition, out of 87 carefully selected living cases in which the state of the heart was noted, and in which the coincidence of symptoms fully justified the diagnosis of contracting kidney, there was evidence of cardiac hypertrophy, with or without valvular disease, in 52,—also as nearly as possible 60 per cent. of the cases.

The following table gives the state of the heart in 807 cases of all forms of Bright's disease collected by Bamberger.

Form of Bright's Disease.	Excentric hypertrophy of the whole heart.	Excentric hypertrophy of the left ventricle.	Simple hypertrophy of the whole heart.	Simple hypertrophy of the left ventricle.	Simple dilatation of the whole heart.	Simple dilatation of the left ventricle.	Total.
Acute	3	4	2	2	4	0	15
Chronic	51	38	3	24	5	1	122
Contracting	88	65	6	39	7	2	207
Total	142	107	11	65	16	3	344

Senator has endéavoured to show that in contracting kidney the heart undergoes concentric hypertrophy, while in fatty kidney the hypertrophy is accompanied by dilatation, causing excentric hypertrophy. Hanot has supported these views. While every pathologist will admit the frequency with which hypertrophy with very little dilatation accompanies contracting kidney, all pathological records disprove such an arbitrary division. Grawitz and Israel found in artificial nephritis, that while the small red or large white kidney resulted indifferently, concentric and excentric cardiac hypertrophy occurred also without any relation to the type of the accompanying change in the kidney, and this is equally true in human pathology.

Traube has the merit of being the first to draw attention to the hard radial pulse so frequently present in contracting kidney; and this fact, together with the accentuation of the second sound of the heart, has been particularly insisted on by Johnson. Sibson drew attention to the doubling of the first sound often accompanying these two phenomena, and explained all three as evidences of increased blood pressure in the aortic system.

Galabin has shown that the high tension pulse occurs in all forms of renal disease, but is most constant in cases of contracting kidney.

Our present purpose is to enquire into the nature and causes of these changes.

This pulse of high tension can be readily recognised. In some cases the radial artery is hard and prominent, feeling, as has been said, like the spermatic cord, but without any unevenness to suggest calcareous deposit; in other cases there is nothing noteworthy about the feel of the vessel. The compressibility of the pulse can be best estimated by placing the forefingers of both hands side by side upon the artery, gradual pressure being made with the proximal finger, while the distal finger notes the effect upon the pulse.

The sphygmographic tracing of the high tension pulse is characteristic.

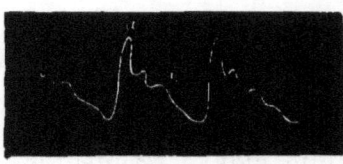

Fig. 4. Normal pulse tracing. *a*, At summit of percussion up-stroke; *b*, Dicrotic notch; *a* to *b*, Tidal wave.

Fig. 4 represents a tracing from a normal pulse. The letter *a* is placed at the summit of the percussion upstroke, which is caused by the sudden impulse of the ventricular systole, and corresponds in height to the force of the left ventricle. The letter *b* marks the dicrotic notch; the little fall is due to the termination of the *systole* and the little rise to the recoil of the vessels and the closure of the aortic valves. From *a* to *b* is the curve of the tidal wave, which lies below the straight line drawn from *a* to *b*. The position of this tidal wave line is the index of tension; it rises with the tension until in high tension pulse tracings it lies above the line *a b*.

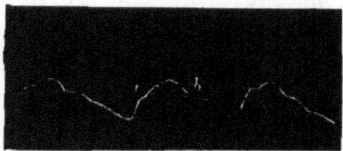

Fig. 5. Pulse tracing, acute nephritis, showing high tension with a relatively feeble ventricle (myocarditis?).

Fig. 6. Pulse tracing; chronic nephritis, with heart failure. High tension with a relatively feeble ventricle (brown atrophy and incompetent aortic valves).

Figs. 5, 6 and 7 are tracings of pulses from three typical cases of Bright's disease. *Fig.* 5 is from acute catarrhal nephritis after sore throat; *Fig.* 6 is from a case of large white kidney which died from heart failure; and *Fig.* 7 is from a well marked case of contracting kidney which led to complete blindness from albuminuric retinitis.

Fig. 7. Pulse tracing; contracting kidney. High tension, with cardiac hypertrophy.

In *Figs.* 5 and 6 the short upstroke is due to the relatively feeble left ventricle. In the acute case there were signs of cardiac dilatation probably caused by acute softening (myocarditis); and in the second case we had *post mortem* evidence of the state of the heart.

Bright, in recording the various organic changes in a hundred cases of renal disease, says, "The obvious structural changes in the heart have consisted chiefly of hypertrophy, with or without valvular disease; and what is most striking, out of fifty-two cases of hypertrophy, no valvular disease whatsoever could be detected in thirty-four; but in eleven of these thirty-four, more or less disease existed in the coats of the aorta; still, however, leaving twenty-two without any probable organic cause for the marked hypertrophy generally affecting the left ventricle. This naturally leads us to look for some local cause for the unusual efforts to which the heart has been impelled; and the two most ready solutions appear to be, either that the altered quality of the blood affords irregular and unwonted stimulus to the organ immediately, or, that it so affects the minute and capillary circulation as to render greater action necessary to force the blood through the distant sub-divisions of the vascular system."

It will be found that the pendulum of modern opinion is now again oscillating between these two explanations in spite of forty years of discussion and observation with the aid which the modern use of the microscope has lent to pathological researches. Thus Senator suggests that in chronic parenchymatous nephritis the hypertrophy is probably attributable to the increased capillary resistance; while, in interstitial nephritis, the hypertrophy is due to direct irritation of the heart, either from some nervous disorder, as in Graves' disease, or, as is more likely, from the blood dyscrasia.

Since the time of Bright there have been four original explanations brought forward and all other writers have adopted some one, or a combination, of these four theories.

Traube, to whom we owe the observation of the hard pulse of Bright's disease, regarded the destruction of a large capillary area in the kidneys as necessarily causing so much obstruction to the circulation that, aided by the imperfect elimination of water, the blood pressure in the aortic system must rise and cardiac hypertrophy follow. Bamberger objected to this that the hypertrophy begins in the earlier stages of Bright's disease; moreover it is present in chronic parenchymatous nephritis in which no destruction of capillaries has occurred. Ludwig and his pupils have shown that ligature of both renal arteries, or of even larger arteries, does not raise the blood pressure in the aorta; while it is well known that in contracting kidney the elimination of water is rather in excess of the normal, yet it is specially in this affection that cardiac hypertrophy occurs. It is, therefore, not without reason that this hypothesis has been generally abandoned, although in recent times it received the support of so eminent an authority as Bartels.

The next original explanation was that given by Johnson, which has undergone at least two modifications. In the first place Johnson pointed out the excessive thickening of the muscular walls of the renal arterioles, and suggested that the obstruction to the circulation offered by a state of tonic spasm in these vessels would explain their own alterations, the rise of blood pressure and the cardiac hypertrophy. But having been able to discover similar changes in the vessels of the pia mater and mesentery, he enlarged his hypothesis and imagined a state of tonic spasm of the whole systemic arterioles which he attributed at first to direct irritation by the impure blood, and later on to stimulation of the vaso-motor centre. According to his latest views he regards the condition as analogous to asphyxia, in which unoxygenated blood going to the brain stimulates the vaso-motor centre in the medulla, and causes contraction of the arterioles throughout the body with consequent increase of the arterial blood pressure. A great objection to this theory of general constriction of the vascular system is that, under such circumstances, the urinary secretion would be diminished or suppressed, as happens in asphyxia (EICHHORST); whereas in contracting kidney, as is well known, the rise in blood pressure is accompanied by an increase in the flow of urine. Grützner has shown that the diuresis excited by the intravascular injection of salts, such as nitrate of potash, is arrested during suspension of the respiration, unless the nerves to the kidney are divided.

The third original explanation is that of Gull and Sutton. They rediscovered the vascular changes described by Johnson, but drew special attention to the thickening of the inner and outer coats, and asserted that the muscular coat was atrophied. They regarded these vascular changes as primary and essential; the increased blood pressure and cardiac hypertrophy are results of them, while the kidney disease is a mere local expression of a generalised degeneration of the arterioles and capillaries, attended by atrophy of adjacent tissues. With regard to these assertions we owe to Thoma the careful measurements which demonstrate that the arterioles of the kidney are absolutely dilated, in spite of the increased thickness of their walls. He showed, too, by careful injections, that while fluids run well into the Malpighian bodies, the efferent artery is often destroyed, and the capillary plexus on the tubules is to a large extent obliterated.

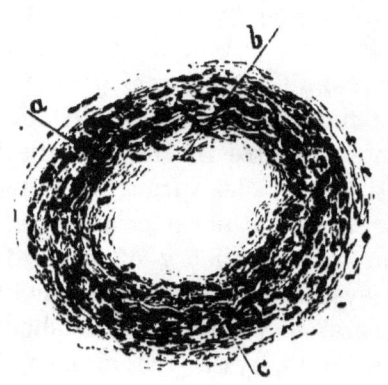

Fig. 8. Renal arteriole shewing Endarteritis obliterans. *a*, Swollen elastic lamina, its fibres separated and œdematous; *b*, Broad growth of connective tissue from the endothelial layer; *c*, Swollen muscular coat with large distinct nuclei (contracting kidney).

Moreover an unprejudiced investigation of sufficient extent will convince any one that many vessels show great hypertrophy of the muscular coat, and that not a few present an appearance of a concentric arrangement of spindle cells lined by swollen elastic tissue. In others, however, the inner coat is much thickened, forming a broad layer of lowly organised connective tissue; the muscular coat may be atrophied, and the adventitia passes inseparably into the surrounding connective tissue. (*Fig.* 8.)

I have looked for the changes in pia mater, peritoneum, skin and other tissues. In the pia mater thickened vessels are often seen, in other situations very rarely. But the pia mater itself is often thickened in chronic Bright's disease, looking opaque, and more or less milky. It is in this condition that the vessels are found thickened, and the true interpretation seems to be that it is a chronic endo- and peri-arteritis associated with sub-inflammatory changes in the perivascular tissues.

But, apart from this, the changes are not constant. Ewald found out of twenty cases of contracting kidney, with hypertrophied heart, the vessels were not thickened in four, while in a similar examination I found them not thickened in two cases out of ten, so that this disposes of the assertion that the vascular lesion is primary and essential. As, too, in chronic nephritis following an acute febrile attack, cardiac hypertrophy is common while the vascular changes are exceptional, they cannot be regarded as standing in any important relation to one another. Finally, we know that both cardiac hypertrophy and high tension pulse occur in acute nephritis, and even in "surgical kidney," in which there is no question of "arterio-capillary fibrosis."

The last explanation is that of Buhl. He believes the cardiac and renal changes proceed *pari passu*, the hypertrophy is to be attributed to the increased activity of the heart. Myocarditis occurs very early, and may lead to no alteration, or to atrophy or hypertrophy of the organ according to circumstances. Most frequently it causes hypertrophy. As the inflammatory process comes to an end the cardiac muscle hypertrophies from excess of nutrition, and to overcome the increased work of the dilated ventricle. In addition he asserts that a relative stenosis of the aorta is present. The rise of blood pressure in the aortic system is therefore due to the hypertrophy of the ventricle and the narrowing of the aorta.

The relative frequency of myocarditis in renal disease is an undoubted fact, and this process plays an important part in many cases, but there is no warrant for the assumption that it has occurred in all cases. Buhl is not at all clear as to the cause upon which the increased activity of the heart or the myocarditis depends. He cannot claim novelty for the fact that the aorta is in some cases relatively stenosed, for Bamberger adduced it as opposed to Traube's hypothesis. With regard to its frequency in granular kidney Ewald has measured the circumference of the aorta above the valves in twelve cases, and found it to vary between 12·1 and 5·7 cm., the average being 7·6 cm., while the normal circumference according to Bouillaud is 6·3 cm., so that very much importance cannot be attached to it as a cause of cardiac hypertrophy.

Having failed to find any of these explanations quite satisfactory, let us return to Bright's suggestions, either that the altered quality of the blood affords irregular and unwonted stimulus to the heart immediately; or, that it so affects the capillary circulation as to render greater action necessary to force the blood through the distant subdivisions of the vascular system. Both these explanations rest upon the supposed impure state of the blood. In the acute and chronic stages of parenchymatous nephritis the condition of the blood was investigated many years ago by Bostock, Gregory, and Christison. According to the latter, urea was always found in considerable quantity in the blood whenever the excretion of urine was diminished; the density of the blood serum was always less, and the fibrine was frequently increased. Nothing of importance has been added to this observation in modern times; but experiments with animals have shown that urea, extractive matters of the blood, creatine and leucin accumulate in large quantities in the blood and tissues after nephrotomy; and it cannot require much argument to convince anyone that during the abeyance of the renal function the blood depurating process must be more or less incomplete.

In contracting kidney the density of the serum is diminished (REES, RAYER); the proportion of salts and albumen is lowered; and there is a rapid reduction of the blood pigment or hæmoglobin (LEICHTENSTEIN). Bartels published some observations, but none of them bearing on the state of the blood in the earlier stages; in the later stages, when dropsy was present, the density of the blood serum was low; a certain amount of urea was found in many cases, in others it was absent.

There are other reasons for believing that blood impurities are early and important phenomena in contracting kidney. Todd first pointed out the frequent co-existence of this form of renal disease with gout, hence the name gouty kidney; Ollivier has drawn attention to its frequent occurrence among workers in lead. Johnson states that the disease is common in persons who "eat and drink to excess, or who, not being intemperate in food or drink suffer from certain forms of dyspepsia, without the complication of gouty paroxysms." He says, that "renal degeneration is probably a consequence of the long-continued elimination of products of faulty digestion through the kidneys." Murchison was persuaded of the relation borne by contracted kidney to persistent lithæmia. Semmola, of Naples, maintains the view that Bright's disease is a consequence of the blood dyscrasia resulting from suppression of the respiratory function of the skin.

Therefore, although during the greater part of its course this affection does not lead to any diminution of the renal secretion, there is ground for believing in a blood dyscrasia depending upon other causes.

The next point for inquiry is whether such blood impurity as may arise from defective elimination or perverted digestive functions can be shown under any conditions to obstruct the capillary circulation. Heidenhain speaks of copious diuresis occurring after the injection of urea *in spite of the blood pressure remaining below normal, or not being proportionately increased.*

Ustimowitsch and Grützner after injecting urea into the blood observed a certain constant rise in blood pressure accompanied by increased flow of urine.

Rendu in his inaugural thesis quotes Potain as having noticed that although injections of urea into the blood do not modify the mechanical conditions of the circulation, yet if a mixture of urea and blood serum be allowed to stand some hours, and be then injected, the arterial tension rises to an unexpected degree.

Grawitz and Israel found that neither unilateral artificial nephritis nor extirpation of one kidney, although followed by hypertrophy of the heart, effected any rise in the blood pressure. By taking strict antiseptic precautions they were able to clamp either of the renal arteries of a rabbit without any unfavourable surgical result. One and a half to two hours afterwards the clamp was removed. In half an hour after the operation the urine became bloody, and in from one to two hours the organ if examined was found to be greyish red and opaque. If the clamp was allowed to remain on longer the kidney became of a dirty yellowish grey colour, indicating commencing gangrene. The effect of this operation was to cause intense acute parenchymatous nephritis with fatty degeneration of the epithelium of the tubules. This passed into either granular atrophy or the large white kidney. In the former, microscopical examination showed no trace of nuclear prolification, the substance of the organ consisting simply of wasted tubules. In another series of experiments they extirpated one kidney altogether. They found that the results of these conditions varied accordingly as the animals operated on were young and growing, or old, strong, and fully grown.

In young animals, after two or three days, the intact kidney began to increase in size to 20 or 30 per cent. more than its normal weight. As the contraction of the other kidney went on, the intact organ continued to increase until it equalled the weight of the two kidneys of an animal of the same size. At the same time they ascertained by abdominal section that the other kidney still secreted a watery urine of low specific gravity, but in small quantity.

In old animals the intact kidney also increased a little, the greatest being in one case of nephrotomy, where the extirpated right kidney weighed 7·7 grms., and the left, after 82 days, weighed 11·3 grms. The hypertrophy of the kidney consisted in a true hyperplasia of the renal elements, at least the enlargement was certainly not due to increase in the diameters of the tubules. The consequence of the imperfect compensation by hypertrophy of the other kidney caused in some cases death by acute or chronic uræmia; in others the animals continued to live, but were ill nourished until the deficit was covered by hypertrophy of the left ventricle.

By careful estimations of the relative normal weights of the heart and kidneys, and comparison of these data with the weights of the altered organs, they were able to determine that the cardiac hypertrophy bore a definite proportion to the loss of renal substance, and therefore was truly compensatory for the renal defect. As a rule the heart was not dilated, in many cases the ventricle was in a state of spasmodic contraction, and they were unable to obtain any confirmation of Senator's views already quoted relative to the pathogenesis of excentric and concentric hypertrophy.

In those cases where cardiac dilatation occurred, the symptoms during life indicated a primarily defective or later destroyed compensation, and the animals died with acute or chronic dropsy. In the case of a large black doe, which died with dropsy of all the serous cavities ten days after the operation, the cardiac muscle was intensely granular, even after the addition of soda solution. They regard this as evidence of the occasional occurrence of *primary* myocarditis. Much more frequently the myocarditis was secondary, but whichever it was it produced the appearance of "excentric hypertrophy." They were never able to obtain evidence of any increased blood pressure, but by careful measurements they determined a constant increase in the velocity of the circulation. Injections of urea also failed to produce a rise of blood pressure, but *stimulated the heart's* action and quickened the blood stream.

More recently Grawitz has shown that cardiac hypertrophy may be induced by loading the blood with urea by giving it to rabbits in their food.

Here we have the explanation of the "self increased activity" of the heart, to which Buhl refers, and also of the myocarditis. The impure state of the blood acts as Bright suggested, by affording an unwonted stimulus to the heart immediately, and this leads to a hypertrophy proportionately compensating for the loss of renal secreting substance. Myocarditis is the cause of dilatation and the subsequent hypertrophy takes place as Buhl indicates, from over nutrition and increased work due to the greater capacity of the ventricle. The rise in blood pressure is not a cause of the cardiac hypertrophy, but a consequence; yet in these experiments the state of the heart, the renal condition, and the blood impurity combined, were unable to effect any increase in the tension of the aortic system.

But the high pressure pulse of Bright's disease is a constant and now universally admitted fact. Traube asserted, not without reason, that he could diagnose contracting kidney by the pulse alone, and Galabin has shown that the same condition is present in the other forms of Bright's disease. A great step was taken when Mahomed proved that this rise in blood pressure precedes the occurrence of albuminuria, the development of which he watched at the termination of scarlatina. Here there is no question of structural change in the heart or arterioles; the sole condition present is that of faulty elimination due to the morbid state of the skin. When constipation occurs the blood pressure rises, and if not averted by a sharp purge, albuminuria follows. Moreover, Mahomed recorded cases of high arterial tension, sometimes accompanied by albuminuria, in young dyspeptic patients free from cardiac hypertrophy.

If the rise in pressure precedes all structural change

it must be due to increased energy in the cardiac contractions, or to obstruction in the distant parts of the vascular system, or to both combined. Most modern writers, Grainger Stewart, Broadbent, Mahomed, Ewald, and others, regard the obstruction as seated in the capillaries, and the cardiac hypertrophy as the consequence of this impediment to the circulation. But we have seen that the cardiac hypertrophy must be regarded as directly dependent on the state of the blood, and therefore is rather to be regarded as a cause than an effect of the rise in blood pressure.

Grawitz and Israel appear to have proved that the cardiac hypertrophy *per se* does not raise the blood pressure, nor does the state of the blood which manifests itself by the stimulus to the heart's action seem to suffice to produce any peripheral obstruction great enough to cause this result. We are compelled therefore to seek some other factor, or to believe that in some respect the conditions of these experiments differ from those of patients suffering from Bright's disease. Such a difference plainly existed in the intact state of one kidney, and it may be that the rise of blood pressure, with which we are clinically familiar, is the result of a dyscrasia differing in some respects from that present in these experiments.

Hamilton has suggested that the increased capillary resistance may be explained by alterations in the specific gravity of the blood serum, and gives the following ingenious explanation of the way in which obstruction would be effected by this change.

Any one who has watched the circulation of the blood in a frog's web or mesentery, has noticed the way in which the corpuscles run in a central core in the blood vessels. "This is due to their being of almost exactly the same specific gravity as the serum. If a body of the same specific gravity as the liquid in which it is suspended is made to travel through either a straight or bent tube, it runs in the axial stream, while, if it is lighter or heavier than the liquid it runs respectively on the upper or lower surface of the tube. The coloured blood corpuscles are almost of exactly the same specific gravity as the *living* plasma in which they are suspended, while the colourless are markedly lighter. Now the whole essence of the blood as a circulating fluid, is that the coloured corpuscles suspended in it never touch the wall of the smaller vessels, even of the minutest capillaries. They glide along in the axial stream without any sign of impediment. A body of the same specific gravity as the liquid in which it is suspended further turns round a curve in the tube with ease, while one lighter or heavier experiences the greatest difficulty in doing so and continually tends to catch at the bends.

"Hence from the great mass of the solid particulate matter of the blood being practically of like specific gravity with the living plasma, undue friction is avoided, and the blood as a whole circulates with very much the same facility as pure plasma would. When a light substance such as oil or milk is introduced into the circulation it will not circulate, because the particles catch in the curves of the vessels, more especially in parts like the lung, where the vessels are particularly tortuous. Blood circulates through a capillary glass tube exactly as it does through a small vein or capillary vessel. The coloured corpuscles occupy the axial stream, the colourless the peripheral. Hence the phenomenon must be a purely physical one. Further, the position in the stream occupied by bodies thus suspended in a liquid may be altered at will, not only by altering the specific gravity of the bodies themselves, but also by changing that of the suspending liquid.

"Thus by substituting a very light liquid or a very

heavy one, the suspended blood corpuscles may be made at will to course along either the lower or the upper surface of the vessel. They come in contact with it, and by rubbing against the wall are made to roll instead of to glide. The friction thus caused retards their progress; they tend like oil globules to catch in the curves of the vessels and thus to cause obstruction to the ever continuous flow of the blood stream. The difficulty that may thus be caused by this apparently trivial cause in moving the blood onwards, comes to be very great when estimated all over the body; and were it the case that a marked difference in the relative specific gravities of the blood plasma and the coloured corpuscles prevailed, the continuance of the even onflow of the blood with the usual propelling power would become a physical impossibility."

I have made this long quotation in order to secure the advantage of the author's own words, as the explanation is exceedingly interesting and ingenious, and has not received the attention it deserves.

Summary.—1. Cardiac hypertrophy is met with in all forms of Bright's disease, including acute nephritis, but is most common in contracting kidney.

2. It is due to: (*a*) stimulation of the heart to increased activity by the presence of non-eliminated poisonous matter, *e.g.*, urea, in the blood; and by, (*b*) increased capillary resistance.

3. Increased capillary resistance may be ascribed to alterations in the density of the blood plasma.

4. The high tension pulse is due to the same causes, viz: increased energy of the heart and greater capillary resistance.

BIBLIOGRAPHY.

BRIGHT (R.). Tabular view of the morbid appearances occurring in one hundred cases with the secretion of albuminous urine. " Guy's Hosp. Reports," Vol. I., p. 380.

BUHL. Ueber Bright's Granularschwund der Nieren und die damit zusammenhängende Herzhypertrophie. "Mitt. aus dem path. Ins. zu München," 1878, p. 38.

DICKINSON (W. H.). The pathology and treatment of Albuminuria. 2nd edition. London, 1877, p. 178.

EICHHORST (H.). Der Einfluss des behinderten Lungengaswechsels beim Menschen auf den Stickstoffgehalt des Harns. " Virchow's Archiv.," Bd. LXX., p. 56.

EWALD (C. A.). Ueber die Veränderungen kleiner Gefässe bei Morbus Brightii und die darauf bezüglichen Theorien. " Virchow's Archiv.," Bd. LXXI., p. 453.

GALABIN (A. L.). The state of the circulation in acute diseases " Guy's Hosp. Reports," 3rd series, Vol. XIX., 1873-4, p. 61.

GRAWITZ (P.) and ISRAEL (O.). Experimentelle Untersuchungen ueber den Zusammenhang zwischen Nierenerkrankung und Herzhypertrophie. " Virchow's Archiv.," Bd. LXXVII., p. 315.

GRÜTZNER. Beiträge zur Physiologie der Harnsecretion. " Pflüger's Archiv.," Bd. XI., p. 370.

GULL and SUTTON. On the pathology of the morbid state commonly called chronic Bright's disease with contracted kidney. " Med. Chir. Trans.," 1872, Vol. LV., p. 273.

HAMILTON (D. J.). Discussion on Albuminuria. " Glasgow Med. Journal," Vol. XXI., 1884, p. 211.

HANOT (V.). Contribution à l'étude de l'hypertrophie concentrique du ventricule gauche dans la néphrite interstitielle. " Arch. Gén. de Méd.," 1878 II., p. 172.

JOHNSON (G.). Diseases of the kidney. London, 1852.

——————— Lectures on Bright's disease. London, 1873.

——————— Lumleian lectures on the Muscular Arterioles. London, 1877.

MAHOMED (F. A.). Clinical aspects of chronic Bright's disease " Guy's Hosp. Reports," 3rd Series, Vol. XXIV., p. 363.

——————— The Etiology of Bright's disease and the prealbuminuric stage. " Med. Chir. Trans," 2nd series, Vol. XXXIX., p. 197.

RENDU. Etudes comparatives des Néphrites chroniques, ' Thèse de Paris," 1878.

SENATOR (H.). Ueber die Beziehungen der Herzhypertrophie zu Nierenleiden. " Virchow's Archiv.," Bd. LXXIII., p. 313.

SIBSON (F.). Harveian lectures on Bright's disease and its treatment. " Brit. Med. Jour.," 1877 I., p. 33.

STEWART (GRAINGER). A practical treatise on Bright's disease of the Kidneys. 2nd edition, 1871, p. 233.

THOMA (R.). Zur Kenntniss der Circulationsstörung in den Nieren bei chronischer interstitieller Nephritis. " Virchow's Archiv.," Bd. LXXI. Heft. 1 and 2.

TRAUBE. Ueber den Zusammenhang von Herz und Nierenkrankheiten. Berlin, 1856.

USTIMOWITSCH. Beiträge zur Theorie der Harnabsonderung. " Arbeit. aus der physio. Anstalt zu Leipzig," 1871, p. 198.

Chapter V.

PATHOLOGY OF POLYURIA.

In the later stages of acute Bright's disease the urinary secretion, which is greatly diminished at the commencement of the attack, often rises to more than the normal amount; in chronic parenchymatous nephritis the daily quantity of urine is generally excessive, while in the contracting kidney Christison wrote truly "no single symptom has appeared to be so invariable or of so much service for indicating the commencement of the disease as the fact of the patient being awakened once or oftener in the night time by the necessity of passing urine. I have scarcely ever known it wanting when any other local symptom existed; frequently it has been present without any other for a great length of time, and it is so remarkable a deviation from the ordinary rule of health that, although it may have been neglected, no individual can fail to recall it when his memory is tasked on the subject by his physician."

This copious urine is after all a poor secretion, often containing less than the normal total solids, and represents solely an increase in urinary water.

The question how this increased outflow of water is caused may not appear at first sight a very practical one, but it has an important bearing on prognosis and treatment, and its theoretical interest is evidenced by the attempts of most authors to give some satisfactory account of it. Thus Bartels gave the following explanation :—

"Observation teaches us that contracting kidneys which have dwindled down to more or less considerable remnants of secreting glandular tissue, do not merely continue to secrete urine, but in the large proportion of cases actually furnish, in the same interval of time, a larger quantity of urine than healthy kidneys would supply. This, however, takes place only so long as the condition of the hypertrophied left ventricle is capable of maintaining the blood pressure in the aortic system at its abnormal height. That the secretory performances of the kidneys depend upon the elevation of the pressure in the arterial system, is proved as distinctly as is possible by physiological experimentation. If the arterial pressure exceeds its normal bounds, it follows of necessity that, *cæteris paribus*, a larger quantity of urinary fluid must be forced through the renal apparatus during the same interval of time than would take place under normal pressure." . . . . "The greater rapidity with which the secretion of the urine is conducted is, at the same time, the cause of its possessing so invariably low a specific gravity, *i.e.*, of its remaining so relatively poor in solid constituents." . . . . "But so soon as the propulsive power of the hypertrophied heart is reduced, in consequence either of some temporary or permanent influence, the abnormally large amount of urine falls off and the abnormally low specific gravity rises."

On the other hand Johnson takes a different view; he says "There is no reason to suppose that high arterial tension has any direct tendency to cause an increased secretion of urine. In cases of contracted kidney the two conditions are associated, but in the early stages of lardaceous kidney the copious secretion of urine occurs without arterial tension. It is probable that in both classes of cases the copious flow of urine is caused by the diuretic influence upon the kidney of some abnormal products in the circulation,—an influence analogous to that of sugar in diabetes." Now it is Johnson's theory that the renal arterioles are contracted; in his own words, "The contraction of the hypertrophied renal arterioles counteracts the injecting force of the strong left ventricle, and thus prevents an afflux of blood in the capillaries of the kidney."

Grützner found that when he injected saltpetre into the blood, the renal nerves on one side only being divided, the urinary secretion was copious, and equal from both kidneys, while there was only a slight rise in the blood pressure; but when the latter was artificially raised by suspension of respiration, the secretion of urine sank in the kidney with its nerves undivided. That is to say, the saltpetre produced its diuretic influence on both kidneys so long as the flow of blood to them was unimpeded, and without the aid of a general rise in the blood pressure; but the moment the arterioles were contracted under the influence of the vaso-motor nerves stimulated into action by the poisonous effect of the carbonic acid upon their centre in the medulla oblongata, the urinary secretion continued only from that kidney whose arterioles, by division of their nerves, were beyond the reach of vaso-motor action.

Newman has suggested that the polyuria of contracting kidney is due to obstruction of the lymphatics.

Physiologists are now agreed that the *amount* of urine depends for the most part upon the blood pressure in the area of the renal artery; this may in turn depend upon (*a*) Systemic causes, by which the general blood pressure is raised, (*b*) Local causes, determining the afflux of blood to the glomeruli.

The last chapter was devoted to the subject of the high arterial tension of Bright's disease, and to proving that this is due to two factors—(1,) The over activity of the heart stimulated by the presence of excrementitious matters (urea, urates, etc.,) in the blood, and (2,) Impeded capillary circulation from alterations in the blood serum. It is unnecessary to repeat the facts and arguments of that chapter. In the general rise of blood pressure there is the first and most important factor in the causation of polyuria, for nothing can be more certain than the clinical fact already alluded to by Bartels, that as in the later stages the heart fails, the polyuria disappears and dropsy sets in.

Temporary plethora produced by abundant drinking is rapidly followed by polyuria.

In the large white kidney, the renal vessels are *dilated* from inflammation, so as to favour the access of a large amount of blood to the Malpighian bodies ; this in part accounts for the polyuria in this form of Bright's disease.

In early cases of contracting kidney there is nothing to oppose the free afflux of blood to the glomeruli, while the progressive destruction of capillary areas beyond them must tend to raise their blood pressure and favour filtration. But in the advanced stages of contracting kidney the renal artery is, according to Thoma, narrower than normal; this is probably counterbalanced by the reduced number of glomeruli to be supplied, and we know from the same observer that the arterioles are dilated, so that the afflux of blood to the glomeruli is very free. We may therefore conclude that there are local conditions which favour the occurrence of high pressure within the glomeruli.

But hydrostatic pressure alone is not concerned ; the epithelial cells covering the glomeruli participate in the process of secretion. It is to this factor that Johnson ascribes the polyuria. Heidenhain found that a copious flow of urine followed the injection of urates into the blood, while the blood pressure remained low ; and urea is said (LANDOIS) to owe its diuretic action to its influence on the epithelium, though Ustimowitsch and Grützner observed the polyuria produced by it to be accompanied by increased arterial tension,—a result we should expect from what we have learnt previously of its action in stimulating the heart's energy. Finally, differences in the permeability of the glomerular walls must also be held accountable for changes in the amount of urine. Thoma found that in very early stages of the contracting kidney, before the microscope shows any changes in the glomerular wall, this structure becomes abnormally permeable. We may assume a similar increased permeability as being probably present, though this has not been proved, in the large white kidney.

Summary. Polyuria is due to the co-operation of these four factors:—

(*a*) Increased general blood pressure; explained in the last chapter.

(*b*) Increased local blood pressure; due to dilatation of afferent vessels and destruction of capillary areas beyond the glomeruli.

(*c*) Increased activity of the glomerular epithelium, stimulated by the presence of urea, urates, etc., in the blood.

(*d*) Increased permeability of the glomerular walls.

BIBLIOGRAPHY.

BARTELS (C.). Diseases of the Kidney. "Ziemssen's Cyclopædia of the Practice of Medicine." Vol. XV. London, 1877.

GRÜTZNER. Beiträge zur Physiologie der Harnsecretion. "Pflüger's Archiv.," Bd. XI., p. 370.

JOHNSON (G.). Lumleian Lectures on the Muscular Arterioles. "Brit. Med. Journal," 1887, I.

LANDOIS (L.). A Text Book of Human Physiology. "Stirling's Translation," p. 572.

NEWMAN (D.). Discussion on Albuminuria. "Glasgow Med. Journal," 1884.

THOMA (R.). Zur Kenntniss der Circulationsstörung in den Nieren bei chronischer interstitieller Nephritis. "Virchow's Archiv.," Bd. LXXI., pp. 42 and 227.

USTIMOWITSCH. Beiträge zur Theorie der Harnabsonderung. "Ludwig's Arbeiten," 1871, p. 198.

Chapter VI.
PATHOLOGY OF URÆMIA.

Uræmia is the name usually given to certain disturbances of the nervous system arising in the course of Bright's disease and in other serious renal disorders.

There are three types under which these phenomena may be grouped. In one the patient lies in a typhoid condition with a dry tongue and feeble pulse, often vomiting, but showing no disturbance of intelligence or special sense, without convulsions or any loss of consciousness that could be called coma. Such cases are met with in bladder diseases and in obstructive suppression of urine. In the more common type there are a moist tongue, headache, sometimes loss of vision, sometimes sudden deafness, or persistent hiccough, vomiting and diarrhœa, persistent dyspnœa, skin eruptions, hyperæsthesia of skin, tremor, twitchings, Jacksonian epilepsy, convulsions and coma. This is the classical type of uræmia associated with Bright's disease.

In yet a third type there may be convulsions and coma, preceded by epigastric pain, with rapid pulse and deep sighing respiration. This resembles that form of coma met with in diabetes, called "Küssmaul's coma," but it has been described by Riess in eight cases of pure anæmia, in five cases of anæmia with renal disease, and in four cases of gastric and hepatic cancer. Senator has observed it in chronic cystitis, gastric cancer, anæmia and atropine poisoning.

Roberts gives "slow panting and laborious breathing" among the symptoms of obstructive suppression of urine. As this type is little known the following case is of interest.

Case 2. Harriet B., aged twenty, was admitted into the General Hospital, on Sept. 6th, 1884, complaining of severe pain in the left hypochondrium which came on suddenly fourteen days before, and had continued ever since. This was attended by vomiting and purging. On the night of admission there was some epistaxis, which did not recur. Ten months ago she had typhoid fever, for which she was an in-patient at the Queen's Hospital, and she stated that she had never been well since. When she was five or six years of age she had scarlatina, but could remember no other illness. On admission she appeared an anæmic girl with an anxious expression of face. She complained of lancinating pains in the region of the spleen. On examination the tenderness was too exquisite to permit of very exact exploration; but there was evidently a resistant mass occupying the whole of the left hypochondrium. The heart, lungs and liver were apparently normal. Her tongue was moist, streaked with brown and white paste, but clean in the centre. Her bowels were relaxed, with about three motions in twenty-four hours. She vomited two or three times daily. Her urine amounted to about twenty ounces daily, it was albuminous, and contained a little pus, but no sugar. The needle of an aspirator passed into the swelling failed to withdraw any fluid. The day after admission the quantity of pus in the urine increased very much, so as to constitute nearly half a column of deposit in the urine glass.

September 9th, at eight a.m., she had a convulsive attack which lasted five minutes, and she afterwards passed into a semi-comatose condition. When seen at 10.25 a.m., she was lying apparently insensible, but on being shouted to opened her eyes and turned her head. Her eyelids were partially open; there were sordes on the teeth and lips. Her pupils did not react to light, but the conjunctivæ were sensitive to touch. Pulse 120. Respirations 24, *deep and sighing*. On auscultation, air could be heard entering freely into the thorax. The nurse said that just before the convulsions she complained of *pain in the stomach*. 4 p.m.—Pulse 132. Respirations 28; noisy; slightly delirious, crying out occasionally. At 6.30 p.m., the house surgeon, Dr. Morrison, saw her, at the request of the house physician's assistant, with the idea of performing venesection. He found her with sighing respiration, pulse small and scarcely perceptible. Every now and then she moaned "Let me die." Temperature 97°. At 6.45 p.m. she was pulseless, and apparently *in articulo mortis*. A vein at the bend of the right elbow was opened and 24 ounces of a solution of sulphate of soda in distilled water injected. The fluid was neutral, sp. gr. 1018, at a temperature of about 100° F. The apparatus consisted of a glass funnel connected with a fine pointed pipette by rubber tubing. This was first filled with fluid, and the pipette was then inserted into the vein and the funnel raised. By this means a steady flow into the venous system took place. Within a few minutes the pulse returned at the wrist and the breathing became deeper. After 24 ounces were injected, the fluid in the end of the pipette was reddened from the blood and fluid mingling; the flow was then stopped and the arm bound up. The patient revived, became conscious and spoke; said she was thirsty and drank some milk. Temperature 98·6°. At 7.45 p.m. about a drachm of foul urinous pus was drawn off by the catheter. She died at 10.25 p.m., about fourteen hours after the convulsion.

During the last thirty-six hours of her life her temperature was taken every hour. Before the convulsive attack it ranged between 98° and 99·5°. At the time of the attack it was 98·5°, it then rose steadily, till at ten a.m. it had reached 100·6°, falling again by twelve to 98°, and as mentioned above, when Dr. Morrison saw her at half-past six the temperature was only 97°. After the operation it rose to 98·6°, and this was maintained at the last recorded time, nine p.m.

At noon on the day of her death, as she had passed no urine since the night, a catheter was introduced into the bladder and about six ounces were drawn off. It was examined the following day by myself. My note taken at the time says: Colour reddish brown, putrid, alkaline, deposits one-third column of pus; sp. gr. 1011. *Gives a deep brown colour with ferric chloride* not diminished by heating. Albumen about one-third column. No sugar. Deposit under the microscope consisted of granular cells, squamous and pyriform epithelium, bacteria and blood corpuscles.

On the following day a careful *post-mortem* examination was made by Dr. C. E. Purslow, who kindly favoured me with the following notes:—

External appearances.—The body was that of a young woman. *Post-mortem* rigidity was well marked in the lower extremities. There was a small incision in the fold of the left elbow, and the mark of a puncture in the left lumbar region. *Head.*—There was a slight increase of the sub-arachnoid fluid, but otherwise the cranial contents were normal. *Thorax.*—The *heart* weighed 9½ oz. The left ventricle was slightly hypertrophied. Both *lungs* were congested and œdematous, and the lung tissue was friable. The *blood* was very watery, with no disposition to clot. (The result of the intra-venous injection resorted to just before death.) *Abdomen.*—The *liver* weighed 43 oz., and appeared normal. The *spleen* weighed 6½ oz., but was otherwise normal. The *stomach* showed signs of chronic catarrh. The right *kidney* weighed 1¼ oz. only; its ureter was patent; its pelvis was dilated and contained a calculus the size of a pea. The medullary and cortical substances were indistinguishable, and together measured only a quarter of an inch in breadth. The capsule stripped off readily.

In the left hypochondrium there was a large mass, on the upper surface of which, and adherent to it, were the duodenum, pancreas, and descending colon. The surface of the mass was purple coloured, and the colon over it also appeared to have blood extravasated under its mesentery and peritonæal investment. On removing the entire mass it weighed 38 oz. At the upper part there was a cavity the size of an orange, containing purulent fluid, which was accidentally ruptured. On section it was composed of an external cyst wall within which was a mass of recent blood clot; inside this was the left kidney in a condition of saccular dilatation. All normal kidney structure was absent. The pelvis contained an irregularly shaped calculus the size of a bean, and another smaller calculus lay in one of the saccules. The *ureter* was dilated but patent. The *bladder* was not enlarged, or its walls hypertrophied. Its mucous membrane was dark coloured, and presented several wart-like outgrowths about one line in height, with rounded surfaces. The *uterus* and *ovaries* were quite normal.

It is very hard to explain these differences in the clinical phenomena of uræmia on the assumption that the simple retention of urea in the blood is the cause of all of them.

When we go a step further and enquire into the various theories that have been propounded to explain these phenomena, we find our difficulties increase.

These theories may be divided into two groups: (*a*) Mechanical (*b*) Chemical. In the first group, Owen Rees, Traube, and Rilliet attributed the nervous phenomena to œdema of the brain, dependent upon the watery condition of the blood and the increased blood pressure; this has been supplemented by Rosenstein, who has suggested that the initial change is spasm of the cerebral blood vessels leading to convulsions by cutting off the blood supply, and followed by effusion of serum into the lymph spaces of the brain. Plainly such a hypothesis involves the necessity for some toxic agent to stimulate the vaso-motor centre, so that it brings us ultimately to some sort of chemical theory; but the fact that the brain is œdematous is disputed by many.

Carter, of Liverpool, in the Bradshawe lecture for 1888, gives the results of his actual determination of the amount of water present in the brains of two patients dying of uræmia. In one case twenty grammes of partly white and partly grey matter were taken from the middle lobe, carefully dried for forty-eight hours at 82° C. over sulphuric acid, then pulverised and dried again in a similar manner until weight was no longer lost. The weight when the drying was complete was 4·15 grammes. The fluid part therefore equalled 15·83 grammes; the percentage proportions being 79·25 water to 20·75 solids, or almost exactly those of normal brain substance, namely, 80 water and 20 solids. In the second case the brain was examined in exactly the same way, and the percentages were 74·55 liquid and 25·45 solids, the water being actually less than in normal brain. The value of these figures is perhaps lessened by the fact that fluid leaves the tissues and drains into the lymph spaces after death, so that fluid which may have been in the brain substance during life may have drained into the ventricles before the examination was made. But Bartels reported several cases of uræmia in which he had noticed absence of cerebral œdema, hypertrophy of the heart, and variations in the density of the blood serum.

The chief chemical theories are :—(1,) That it is due to urea in the blood, hence the name uræmia, originally given to it by Piorry, and supported by Christison. This theory is the one which has managed to hold its ground, although numerous experimenters have failed to produce any toxic phenomena by the injection of urea into the blood of animals or by making them ingest quantities of urea with their food.

Peabody has calculated on data obtained by experiments on dogs that at least 1½ lbs. of urea would be required to prove fatal, whereas in the body of a man dying of uræmia only ·009 lb. could be recovered. According to Gréhant and Quinquaud the quantity of urea required to produce convulsions in dogs is from $\frac{1}{100}$th to $\frac{1}{50}$th of their body weight, confirming Peabody's estimate.

In uræmia the amount of urea found in the blood has varied from ·2 to 1 per cent. Vierordt estimates the total quantity of blood in man to be 5062·5 grammes or about $\frac{1}{13}$th of the body weight, so that even the highest estimate, 1 per cent., would only give about 1½ oz. instead of the requisite minimum of 1½ lbs.

Snyers found that he could inject into a dog doses of urea equivalent to the amount it would eliminate in three days without producing any ill effects.

Cases have been described by Owen Rees, Christison, Bright, and Frerichs, in which the blood contained large quantities of urea without giving rise to any symptoms of uræmia. Biermer has published a case of anuria lasting 118 days without uræmic phenomena. These appeared after the urine began to flow.

In the following case uræmic symptoms persisted in spite of the elimination of a normal quantity of urea; similar cases have been reported by Rosenstein, Christison, and Liebermeister.

CASE 3. Emily M——, aged 28, was admitted August 7th, 1888, with headache, vomiting, and hæmaturia. She had no dropsy, but her urine, which was persistently albuminous, contained numerous epithelial and hyaline casts. She suffered constantly from frontal headache, and had frequent cramps and vomiting. The urine generally averaged 35 oz. daily. On a diet of fish, chicken, milk, bread, butter and tea, she still complained of these uræmic symptoms, but the quantity of urea eliminated was 450 grains in 24 hours, the percentage being as high as 3·2.

(2.) This urea theory was slightly modified by Frerichs and Treitz, who suggested that the urea, itself innocuous, became converted, under the influence of a peculiar organised ferment, into carbonate of ammonia.

This theory is based upon (a,) The facility with which this transformation is effected ; (b,) The resemblance of the symptoms produced by the intravenous injection of carbonate of ammonia to those of uræmia ; (c,) The presence of carbonate of ammonia in the blood of uræmic patients.

Researches made to determine the point whether carbonate of ammonia is really present in the blood in uræmia have attained very contradictory results. Snyers states that the blood of dogs some days after the ligature of both ureters contains only traces of ammonia.

Rommelaere concludes that the quantity of ammonia in the blood is increased after nephrectomy ; that the quantity is too small to explain the occurrence of uræmia ; but that in a few hours after death a large quantity of ammonia is formed in the blood.

(3.) Schottin in 1853 suggested that the extractives of the blood,—creatin, creatinin, leucin and tyrosin were the real poisons.

In support of this Oppler found a great excess of creatin in the blood of nephrectomised animals, and Hoppe-Seyler found a great accumulation of extractive matter in the blood of a patient who had presented uræmic symptoms in the course of an attack of cholera.

(4.) Gauthier has suggested that ptomaines may be the poisonous substances.

(5.) Feltz and Ritter in 1881 commenced a new series of researches which led to unexpected results. They determined (a,) That the intravenous injection of fresh urine causes convulsions, coma and death; (*b*,) That their results were independent of the increased pressure produced by the injection or of the organic constituents; (*c*,) That the inorganic constituents injected separately produced the same symptoms as the urine itself, and that of these the potassium salts showed the most powerful toxic action.

Astaschewsky published about the same time experiments supporting these conclusions.

There is no doubt of the toxic properties of potassium salts, but Bouchard and Snyers believe that they do not play so exclusive and preponderating a part as Feltz and Ritter affirm.

In the blood of two eclamptic women in Braun's wards at Vienna, Snyers found in one 2·07 parts of potash per thousand parts of urine, and in the other 2·06 per thousand, a quantity rather less than the normal amount.

On the other hand, Lépine found in some experiments on animals by temporary compression of the renal artery, that the urine secreted by the injured kidney showed diminution of its solids chiefly in phosphoric acid and potash, while the chlorides were not diminished, but rather increased.

(6.) Bouchard accepts the position that the urine is itself poisonous, but attributes this to various sources: (*a*,) Food derivatives, especially potash salts; (*b*,) Products of intestinal putrefaction absorbed with the blood: (*c*,) Admixture of bile, saliva, and other secretions; (*d*,) Products of tissue metamorphosis.

Voit says, "Symptoms of disease originate wherever any substance which does not belong to the economy accumulates within the body and is not eliminated from it;" and he shows that even sulphate of soda may be deleterious under such circumstances. The toxic effects are the result of interference with the normal exchanges which take place between the blood and the tissues, and upon which the vital phenomena of the latter depend.

(7.) Lépine has shown that febrile urine is very much more poisonous than non-febrile urine; and Bouchard observed that the day urine is more poisonous than that of the night. The urine in certain diseases is especially poisonous, *e.g.*, pernicious anæmia, jaundice, cancerous cachexia, chronic Bright's disease, &c.

We are not in a position at present to explain precisely the pathogenesis of so-called Uræmia. It is plain that the clinical phenomena vary, and that there are many poisons to which these symptoms may be due.

Since the case of Harriet B. was published (1884), the doctrine that Küssmaul's coma is due to the presence of an acid in the blood has gained ground; it is extremely probable that some cases of uræmia are due to this cause, although the acid has not yet been identified. Carter, in his Bradshawe lecture, relates a case of coma with sighing respiration and "decidedly acid" blood. The acid is probably formed in the intestine, and thence absorbed into the blood.

If we are to get a nearer knowledge of these problems, it must be by differentiating the clinical types, and by recognising the probability that different agencies may be at work in each; moreover in one we may see the effects of a sudden large dose, in another of chronic intoxication.

Summary.—(1,) Uræmia is a convenient generic name given to a large series of nervous accidents which occurs in Bright's disease.

(2,) Its causes must be looked for in certain toxic agencies arising in the blood, in the tissues, or in the intestines, which have not at present been identified with certainty.

(3,) Normal urine contains a certain proportion of some of these poisons, so that its retention is liable to cause them to accumulate in the blood. In disease, these or other poisons may be present in greater quantities with proportionately more serious results.

(4,) Poisons are known to be formed in the intestines; when constipation is present this process is facilitated, and opportunity is given for their absorption into the blood. Moreover, the intestine forms a channel by which elimination may take place when the renal function is depressed, so that failure of intestinal action is a grave additional danger in Bright's disease.

BIBLIOGRAPHY.

ASTASCHEWSKY. Zur Frage von der Urämie. " St. Petersburger Med. Woch.," 1881, No. 27.

BIERMER. Ein ungewöhnlicher Fall von Scharlach. " Virchow's Archiv.," Bd. XIX., p. 537.

BOUCHARD. Poisons de l'organisme et toxicité urinaire. " Compt. Rend.," 6 Juillet, 1885.

BRIGHT (R.). Renal disease accompanied with the secretion of albuminous urine. Case 19.—Urea in the serous fluids. " Guy's Hosp. Reports," Vol. V., 1840, p. 138.

CARTER (W.). The Bradshawe lecture on Uræmia. "Lancet," 1888, II., p. 355

CHRISTISON (R.). On granular degeneration of the Kidnies. Edinb., 1839.

FELTZ and RITTER. De l'urémie expérimentale. Paris, 1881.

FRERICHS (F. T.). Die Bright'sche Nierenkrankheiten und deren Behandlung. Braunschweig, 1851.

GAUTHIER (A.). Sur les ptomaïnes. " Le Progrès Médical," 1886, p. 69. Bull. de la Soc. de Médecine, Jan., 1886.

GRÉHANT and QUINQUAUD. L'urée est un poison; mesure de la dose toxique dans le sang. " Comp. Rend.," Vol. XCIX., 1884, p. 383.

LÉPINE and AUBERT. Sur la toxicité des matières organiques et salines de l'urine normale et fébrile. " Lyon Méd.," 1885, I., p. 101.

OPPLER. Zur Lehre von der Urämie. " Virchow's Archiv.," Bd. XXI,, p. 260.

PEABODY (G. L.). The relation existing between retention of urea and uræmia. " New York Med. Rec.," Vol. XXVII., 1885, p. 22.

REES (OWEN). On the nature and treatment of diseases of the Kidney. London, 1850.

RIESS (L.). Ueber das Vorkommen eines dem sogenannten Coma diabeticum gleichen symptomencomplexes ohne Diabetes. "Zeitsch. für. Klin. Med.," Vol. VII., 1883, Supp. Heft., p. 34.

RILLIET and BARTHEZ. Maladies des Enfants, Vol. I. Paris, 1843, p. 777.

ROBERTS (W.). A practical treatise on Urinary and Renal Diseases. Fourth Ed., p. 34. London: Smith, Elder and Co.

ROMMELAERE ——, quoted by Snyers, op. cit., p. 143.

ROSENSTEIN (A.). Die Pathologie und Therapie der Nierenkrankheiten. 3rd Ed., Berlin, 1886.

SCHOTTIN. Beiträge zur Characteristik der Urämie. "Vierordt's Archiv.," 1853, Heft 1.

—————— "Archiv. der Heilkunde," 1860.

SENATOR (H.). Ueber Selbstinfection durch abnorme zersetzungsvorgänge und ein dadurch bedingtes (dyscrasisches) Coma Küssmaul'scher symptomencomplex des "diabetischen Coma." "Zeitsch. für. Klin. Med.," Bd. VII., 1883, Heft 3, p. 235.

SNYERS (P.). Pathologie des Néphrites chroniques. Bruxelles, 1886.

TRAUBE (L.). Gesamellte Beiträge zur Pathologie und Physiologie. Bd. II., 1871, p. 551.

TREITZ. Ueber urämische Darmaffectionen. "Prag. Viertel.," 1859, Bd. 4.

Chapter VII.
RETINAL CHANGES.

Amaurosis, or defect of vision, was first noticed in association with dropsy by Wells, in 1812; but the first actual observation of retinal change in connection with Bright's disease was made *post mortem* by Türck, in 1850.

The visual defects associated with Bright's disease are divided into two classes: (1,) Those due to uræmic poisoning, in which the visual centres in the brain are chiefly at fault; and (2,) Those due to structural changes in the optic nerve and its retinal expansion.

The so-called uræmic blindness of the former class is generally unattended by any changes in the fundi; but Dobrowolsky has observed transitory œdema of the discs, while Litten has noticed that in uræmic attacks there is swelling and cloudiness about the disc.

In the second class the changes observed in the retina may be enumerated as follow: (1,) Diffused opacity from œdema; (2,) White patches; (3,) Hæmorrhages; (4,) Optic papillitis; (5,) Diffused retinitis, in which many of these may be combined; and (6,) Atrophic changes sequential to inflammation.

Apart from the slight changes which occur in acute uræmic poisoning, no retinal affections are met with in primary acute nephritis, though they are liable to supervene rapidly in the acute attacks which so commonly occur in the course of chronic Bright's disease.

They are most frequent and characteristic in the contracting form, and it is to this that most of the recorded observations refer. It has been disputed whether they ever occur in the lardaceous form, and this point has been settled in the affirmative; but as what is often called lardaceous disease is merely chronic

Bright's disease—*plus* lardaceous degeneration—it is doubtful if the observations recorded are of much value. The existence of retinal changes, in a pure case of primary lardaceous degeneration, has yet to be placed on record.

Nine years ago I collected a hundred cases of contracting kidney from among my out-patients, and these were examined by Mr. Eales, surgeon to the Birmingham and Midland Counties Eye Hospital, who published the result of his observations in the Birmingham Medical Review for January, 1880. Out of the hundred cases, retinal changes were present in twenty-eight, rather less than the number recorded by Galezowski, (fifty out of a hundred and fifty); but decidedly greater than the nine per cent. found by Wagner.

In sixteen of these twenty-eight cases, changes were found in *one* eye only; probably because they are apt to begin in one eye before the other; thus in one case at the first examination no ophthalmoscopic changes were found; two months later several white specks were observed in one eye along a branch of the retinal artery, but there was no evidence of inflammation.

In these sixteen cases the lesions were as follow: In six, several round white patches; in five, one or two spots only; in one, a single recent hæmorrhage near the disc; in two, black specks, associated in one instance with white specks; in one, two large, round, soft edged, whitish patches close to the disc.

In the twelve cases in which both eyes were affected the following lesions were noted: In four, diffuse retinitis in both eyes; in one, diffuse retinitis in one eye, with a single hæmorrhage in the other; in five, many whitish round patches; in two, a few white patches.

In addition to these twenty-eight cases, there were three in which the disc was abnormal, this structure being abnormally pink (hyperæmic) in one, and abnormally pale (atrophic), with blurred edges, in two. The first was probably a case of incipient neuritis, and the others were very likely atrophy secondary to a slight neuritis; so that if we include them we obtain thirty-one cases out of a hundred, a number very close to the thirty-three per cent. recorded by Galezowski.

Although there is no evidence that these retinal changes ever occur in acute nephritis supervening in healthy kidneys, they are not very uncommon in connection with the albuminuria of pregnancy. But the albuminuria of pregnancy is preceded and caused by chronic blood poisoning, of which the retinal changes are only another local expression. Doubtless when the kidneys begin to fail in their function, the blood poisoning rapidly gets worse; but it is a condition which must be dated back some time anterior to the earliest expression of renal trouble.

The various forms of retinal change already enumerated may be classified as (1,) Neuritic or inflammatory; (2,) Hæmorrhagic; (3,) Degenerative.

The manner of their occurrence is well illustrated by the following case:—

CASE 4. William S., aged twenty-three, was admitted on February 24th, 1888, complaining of pains in the head and back and dimness of sight.

History.—About a month ago he noticed that his eyes were getting dim; these continued to get worse until he went to the Eye Hospital on the 23rd, and was sent on from there to the General Hospital. He had never had dropsy, scarlatina or rheumatic fever; nor could he recollect ever being laid up with illness. For two years he had been a teetotaler, but had previously only been a very moderate drinker. He did not use lead in his work. He had suffered from headache for ten years, and was often giddy, but had never had a fit. On two occasions he had lately noticed a little swelling of his ankles, and for the last month he had had to get out of bed to pass water.

Present condition.—There was a greenish line on the margin of the gums, but no blue line. The heart was a little enlarged, with reduplicated 1st sound at apex and accentuated 2nd sound in the aortic area. The pulse was regular, full and incompressible.

The urine measured 78 ozs.; sp. gr. 1010, acid, pale straw colour; 0·8 % of urea; 2·5 grammes per litre of albumen; depositing numerous granular and hyaline casts, white and red blood corpuscles, and renal epithelium. Ophthalmoscopic examination showed diffuse neuro-retinitis in both eyes, but in the right the disc was surrounded by radiating flame-shaped hæmorrhages, while in the left the disc was more swollen, and surrounded by soft white patches of inflammatory exudation. (*Fig.* 9).

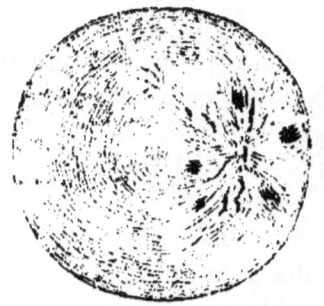 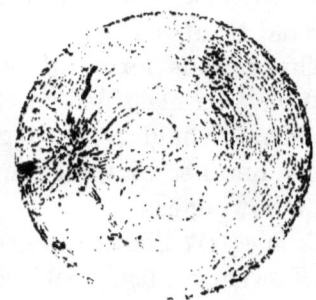

Fig. 9. Right Eye.—Disc swollen and infiltrated; flame shaped hæmorrhages radiating round it.

Left Eye.—Disc much swollen and infiltrated; soft rounded patches of inflammatory exudation round it.

Progress of Case.—He made no improvement, but it is noteworthy that on milk diet he passed a fair amount of urea; thus, on March 7th he passed 317 grains, and on March 16th 374 grains, in twenty-four hours.

On March 26th the eyes were re-examined, the right eye being no better; the swelling was as great, but there were fewer hæmorrhages; in the left eye the swelling and soft patches around the disc had disappeared, while a radiating group of atrophic patches had made their appearance around the yellow spot. (*Fig.* 10).

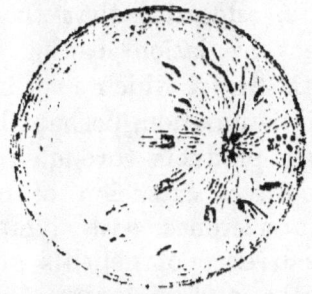
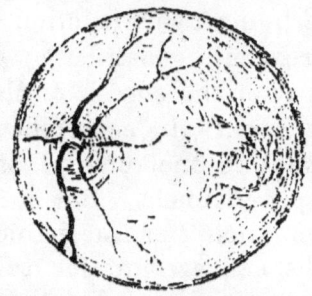

Fig. 10. *Right Eye.*—Swelling of disc undiminished. Hæmorrhages undergoing absorption.

Left Eye.—Disc much less swollen; vessels visible; rounded patches gone. Patches of retinal degeneration radiating round the yellow spot.

He died of cerebral hæmorrhage on April 7th, when his kidneys were found to be in an advanced stage of contraction.

Neuritis and *neuro-retinitis* are not solely dependent upon kidney disease for their causation. They are common in encephalic diseases, especially cerebral tumours, and they are met with in acute diseases of the spinal cord, in diabetes, in lead poisoning, chronic alcoholism, anæmia, (both simple and pernicious,) in measles, scarlatina, typhoid fever, malarial fever, and in sudden suppression of menstruation, while neuritis probably occurs as an idiopathic affection, especially in hypermetropic eyes.

In some of these conditions, notably in certain cases of brain tumour and of anæmia, the ophthalmoscopic appearances are indistinguishable from those of Bright's disease.

The occurrence of optic neuritis in connection with cerebral tumours has been explained by (1,) Assuming a direct propagation of an inflammation starting from the neighbourhood of the tumour; (2,) Increased intracranial pressure causing obstruction to the return of blood from the eye; (3,) The tumour giving rise to irritation as a "foreign body." The last theory is only worth mentioning because it is that of Dr. Hughlings Jackson.

There seems to be good reason to believe that descending neuritis often occurs, and also that there is frequently obstruction of the circulation leading to dropsy of the sheath of the optic nerve, which in some way favours the occurrence of inflammation, perhaps by preventing the removal of waste products through the lymph channels. But neither direct extension of inflammation nor mechanical interference with lymph paths can account for the occurrence of neuritis and neuroretinitis in Bright's disease, and in many other of the pathological states mentioned above.

The condition which is common to all of these is a dyscrasia, or disordered state of the blood, which leads frequently to inflammatory changes in other structures throughout the body.

In Bright's disease the dyscrasia is so marked, and its influence in setting up such inflammatory changes so well recognised, that it must be allowed to afford a very sufficient explanation of the neuro-retinitis.

The appearances seen vary from hyperæmia, with slight œdema of the disc, to swelling from exudation, which hides the disc and its vessels and extends far into the retina, and is often associated with radiating hæmorrhages. The inflammation may subside, and leave atrophy of the disc and retina.

Retinal hæmorrhages in Bright's disease are usually flame shaped, striated patches situated along the course of the vessels radiating from the disc.

They present, however, variations in shape, size and position, as might be expected.

Like neuritis, retinal hæmorrhages are no special appanage of Bright's disease. They are met with in various diseases, and especially in blood disorders, *e.g.*, malaria, scurvy, purpura, leucocythæmia, septicæmia pernicious anæmia, &c.

Their pathology is essentially the same as that of neuritis; that is to say, they depend upon a dyscrasia, but in Bright's disease there are three factors which must be allowed to share in their production, these are (1,) The dyscrasia; (2,) The high arterial tension; and (3,) Degenerative changes in the small vessels. Gowers has described irregular dilatation of capillaries in Bright's disease, probably resulting from changes in their walls, and it is at least probable that some of the hæmorrhages are due to the breaking of the vessels at such weakened parts. Aneurisms of the small arteries have also been observed, but they are rare.

White patches of degeneration are met with under two sets of conditions; in one they are primary, in the other secondary to hæmorrhages and inflammation. In the primary form the patches are small at first, and are usually in the neighbourhood of the yellow spot. They are often arranged so as to form streaks radiating from it (see *Fig.* 10). Quite similar appearances have been seen in diabetes.

In course of time they may coalesce and form larger irregularly shaped areas.

The secondary variety occurs chiefly around the disc. They are generally larger and more irregular in shape, though their outline forms a series of curves (see *Fig.* 9).

The patches of primary degeneration must be attributed to the dyscrasia, which gives rise to so many analogous changes in other tissues of the body. The process is essentially a retrograde metamorphosis of the highly differentiated retinal tissues into undifferentiated connective tissue.

Where the change is secondary to inflammation, compound granule cells and other evidences of the fatty degeneration of an inflammatory exudation are seen under the microscope. (*Fig.* 11).

In connection with these sequential changes Brailey and Edmunds find that the retinal vessels are always thickened by endarteritis, which may go on to obliteration of the lumen; this change may precede any alteration recognisable by the ophthalmoscope, but the same remark applies that has been made of this condition elsewhere, namely, that it is part and parcel of inflammation of the surrounding tissues, either present or past.

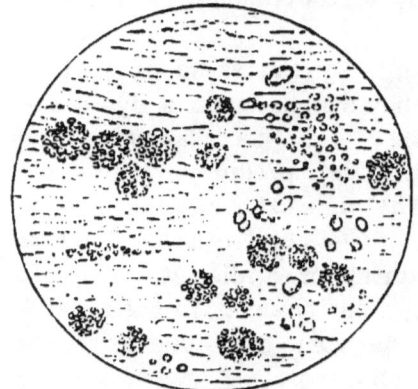

Fig. 11. Section through atrophic retinal patch shewing fatty degeneration of exudation. × 250. (after GOWERS).

Narrowing of the arteries is very common, associated with dilatation of the veins. Gowers believes in a persistent spasm of the arteries, but this narrowing is constantly seen as a consequence of retinal disease apart from Bright's disease.

In reference to this question Eales says, "I cannot confirm the statement of Dr. Gowers, that this (arterial contraction) is common. Only twice, I think, some slight contraction existed, but I did not feel sure that it was abnormal even in these cases.

"I have recently had a case of diffuse hæmorrhage into the retina; the hæmorrhages have cleared; the retina looks quite healthy, but vision is imperfect, and the arteries have contracted very much since I first saw the case. Here, though no sign of disease is visible, I think we may safely infer that the retina is damaged, and that the contraction is a consequence of degeneration in the retina, as it was not noticeable in the early stage of this case, and contraction of the retinal arteries from this cause is common."

Choroidal hæmorrhage may occur leading to circumscribed atrophy and pigmentary disturbance.

"Colloid" degeneration of the vessels of the choroid has been described by Poncet.

After parenchymatous retinitis there may be some pigmentary degeneration of the choroid in the form of small grey spots arranged in groups (GOWERS).

Thickening of the adventitia or lymph sheath is very common, but always in association with retinal disease.

Embolism of retinal arteries has been described, but of late years it has come to be recognised that plugging of the vessels is more usually thrombotic, and it is probable that these statements should be taken to imply simply that the artery was blocked by a clot, which blocking, in accordance with the prevalent doctrines of that time, was assumed to have come from a distance instead of being formed *in situ*, as is now recognised to be the more probable explanation.

Hæmorrhage into the vitreous may occur from the bursting of a large retinal extravasation. It causes permanent damage to vision and may give rise to glaucoma (GOWERS).

Detachment of the retina may occur occasionally, as the result of serous effusion between the retina and the choroid.

An example of this rare condition has been reported by Anderson.

CASE 5. The patient was a girl aged nine years, admitted into the London Hospital on March 14th, 1887, under the care of Dr. Samuel Fenwick. She complained of sickness and headache, was extremely pale and wasted, but showed no œdema of face or limbs. She passed 50 to 65 ounces of urine in twenty-four hours, 1010 to 1012 specific gravity, with one-fourth albumen, some free blood discs and casts of various kinds. The heart was hypertrophied and the arteria tension high. The opthalmoscope showed double neuro-retinitis. The child had been healthy until eighteen months of age, then had an attack of measles, and was never subsequently well. In November, 1886, she was noticed to have frequent nocturnal micturition, and five weeks before admission she complained of headache and sickness, and that she could not thread a needle. A fortnight before admission into the hospital she had a severe fit, was universally convulsed, and then lay unconscious for three days. When she recovered consciousness she was practically blind, but partly recovered vision.

Dr. Anderson first saw the child on April 10th, when he found severe neuro-retinitis with numerous pale hæmorrhages, and considerable pale exudation in the papilla and retina. On the nasal side of each fundus there was extensive detachment of the retina, which was greyish pink and glistening, with the vessels to be seen climbing over it. The surface of the detachments oscillated freely but slowly when the head was moved. The child was almost quite blind, but mentally clear. The retinal detachments rapidly increased. That of the left eye showed numerous bladder-like bulgings round the lower and nasal periphery. That of the right eye showed four large detachments almost meeting in the middle, leaving only a narrow quadrangular chink, at the bottom of which the fundus could be seen. The child rapidly got worse, the face and limbs got slightly puffy, the urine diminished and became almost pure blood; she became drowsy and died comatose on April 24th, three months after the first complaint of visual defect. The necropsy showed advanced fibroid contraction of the kidneys, the left weighing only three quarters of an ounce and the right two ounces and a half. The retinæ were separated from the choroid by clear straw-coloured fluid.

Effects on vision.—In the great majority of cases presenting retinal change, that is, in those in which only a few degenerative specks or hæmorrhages are found, vision is unaffected. Acute neuro-retinitis may be present without causing any complaint from the patient, but this is exceptional. As a rule vision is affected in proportion to the extent of the disease, but the patients rarely become quite blind, generally retaining perception of light sufficient to enable them to get about if they are otherwise able to do so.

Diagnostic value.—It is abundantly evident that the value of these changes is not very high for purposes of diagnosis, as they are only present in about one in three cases of chronic kidney disease, and generally in the later stages when the diagnosis has been made by other means. There are certain cases in which the eye symptoms are the first to cause the patient to seek medical advice, of which those quoted in this chapter are examples.

As already stated, the ophthalmoscopic appearances are not in themselves pathognomonic.

Cases of diffuse neuro-retinitis simulating retinitis albuminurica have been reported by the best observers (HUGHLINGS JACKSON, GOWERS, EALES); while the hæmorrhages and white patches are seen in various blood diseases.

Prognostic value.—On the other hand their great value for prognosis does not admit of doubt.

Mere white specks are not of any grave significance; they are commonly seen in the albuminuria of pregnancy, and were found in a certain proportion (five out of fourteen) of cases of so-called functional albuminuria examined for me by Mr. Eales.

But in the presence of chronic renal disease, it is almost certain that a grave prognosis must be given. Dr. Miles Miley has made a careful enquiry into the duration of life of cases with albuminuric retinitis in the London Hospital. Of 156 cases of chronic renal disease in which the eyes were examined, 105 were stated to be normal, of these 28 died in hospital; of 51 cases in which the eyes were affected 25 died in hospital; that is 26 per cent. of the first series and 52 per cent. of the second. Out of the whole 51, 45 were known to be dead at the time of publication, the other 6 could not be traced, and out of all those who had died, one lived eighteen months, two fourteen months, and the remainder less than twelve months.

In the discussion at the Ophthalmological Society which followed the reading of this paper, the gravity of the prognosis was generally admitted, but the existence of exceptional cases was referred to, and a case was mentioned, quoted by Dr. Webster, at the American Ophthalmoscopic Society, of a clergyman still living, in whose fundi this condition was recognised ten or fifteen years before. This case is of doubtful value, unless the diagnosis of Bright's disease was established by other symptoms.

The diagnosis of the following case is not free from doubt, but it is worth placing on record as an example of the sort of case that may be occasionally met with presenting apparent exceptions to the grave prognosis which is undoubtedly the rule.

Case 6. Kate ——, aged 19, applied to Mr. Eales on June 23rd, 1881, for dimness of sight, specks before the eyes and headache of eight weeks' duration. Ophthalmoscopic examination revealed appearances resembling albuminuric retinitis in both eyes; viz., there was slight swelling of the disc and retina associated with white soft edged patches around both disc and macula, but without any radiating bright streaks near the latter. The urine showed a sp. gr. of 1010, was albuminous, and contained casts.

She was a rather anæmic girl, and suffered from constipation. The treatment pursued was based on the supposition that she had Bright's disease.

Vision greatly improved; in September $V = \frac{18}{xx}$ in each eye, and there was considerable subsidence of the retinitis.

A year later she had a recurrence of the retinitis chiefly in the right eye in which $V = \frac{15}{v}$, while in the left $V = \frac{15}{xx}$. But this soon subsided, and in February, 1883, $V = \frac{15}{xv}$ in each eye, and up to the present time she has had no recurrence of the retinitis.

Mr. Eales has seen her from time to time, and in October, 1888, he examined her and found $V = \frac{15}{xii}$ in each eye. Ophthalmoscopic signs normal, except that in each fundus there was a small pigment spot in the retina with slight disturbance of the choroidal epithelium. The urine throughout these years has been always albuminous, and casts have been repeatedly found.

For this part of the case I am responsible, but unfortunately my urine reports, which were written on separate pieces of paper, have got mislaid, so I can give no further details.

Three years ago she had a fit, after which she states that she was blind and unconscious for three days. I have made enquiries from her medical attendant as to his impression of the nature of the attack, but he cannot recollect. At the present time she is an anæmic-looking girl, who suffers from dyspepsia and constipation; there is no dropsy, nor any sign of high arterial tension or organic disease. She menstruates regularly; the amount is sometimes scanty, never excessive. There is a hæmic murmur at the base of the heart. The urine was clear, pale, acid, sp. gr. 1008; contained a good cloud of albumen; deposited epithelium and oxalates but no casts.

We shall continue to watch the case with interest.

The occurrence of diffuse retinitis in the albuminuria of pregnancy, well-known to be a condition admitting of a favourable prognosis, is an apparent exception; but as already pointed out, the eye symptoms here depend upon a dyscrasia not solely resulting from the kidney disease, which is often slight, and usually curable, while the dyscrasia improves after parturition.

Possibilities of cure.—Though all authorities admit the possibility of cure and the disappearance of spots, and instances of marked improvement have been already related, it is doubtful whether except in association with the albuminuria of pregnancy recovery ever does take place. This is due, not to any essential incurability in the retinal condition, but to its dependence upon an incurable renal disease in its last stages. From what has been said of the general prognosis of cases in which these retinal changes are present, it is sufficiently obvious that there is little room to discuss their curability.

Summary. 1.—The chief retinal changes of Bright's disease consist of neuritis and neuro-retinitis, hæmorrhages and white patches of degeneration.

2.—These do not possess any specific characters which enable the observer to diagnose Bright's disease with absolute certainty from them alone.

3.—They are dependent chiefly upon the disordered state of the blood, with which, in the case of hæmorrhages, the high arterial tension and the diseased vascular walls co-operate.

4.—They are seen usually in advanced cases, and are of grave prognosis; the apparent exception of the albuminuria of pregnancy is explained by their being due, in that case, to a dyscrasia not dependent solely upon the renal disease.

BIBLIOGRAPHY.

ANDERSON (J.). A case of subretinal effusion in chronic Nephritis in a Child. "Brit. Med. Jour.," 1888, I., p. 248.

BRAILEY and EDMUNDS. The relation of the retinal changes to the other pathological conditions of Bright's disease. "Ophth. Soc. Trans.," Vol. I., 1881, p. 44.

DOBROWOLSKY. Diffuse Netzhautentzundung bei hochgrädiger Hypermetropie. "Klin. Monatsbl. für Augenheilkunde," Bd. XIX.

EALES (H.). The state of the Retina in one hundred cases of Granular Kidney. "Birm. Med. Review," Vol. IX., 1880, p. 84.

GALEZOWSKI (X.). Traité iconographique d'Ophthalmoscopie. Paris, 1876.

————. De la Rétinite et de la Rétino-choroïdite albuminurique : leur traitement. "L'Union Médicale," 1878, p. 924.

GOWERS (W. R.). A Manual and Atlas of Medical Ophthalmoscopy. 2nd Ed. London, 1882.

JACKSON (J. HUGHLINGS). Case of double optic Neuritis without cerebral Tumour. "Moorfields Hosp. Repts.," Vol. VIII., part 3.

MILEY (MILES). On the prognosis of Neuro-retinitis in Bright's disease. "Brit. Med. Jour.," 1888, I., p. 248.

PONCET. De la Rétinite albuminurique. "Gaz. Med. de Paris," No. 32.

TÜRCK. Anatomisches Befund von Amaurose. "Zeitschr. der Wiener Aerzte," No. 4, 1850.

WELLS (W. C.). Observations on the Dropsy which succeeds Scarlet Fever. "Trans. of a Soc. for the Improvement of Med. and Chir. Knowledge," Vol. III., 1812, p. 167.

Section II.—

CLINICAL EXAMINATION OF THE URINE.

Chapter VIII.

INSPECTION of the urine is an ancient medical practice, which, for the most part, was until recent times a meaningless and perfunctory medical ceremony.

Healthy urine is a clear, bright yellow fluid, having a not disagreeable but characteristic odour. It is liable to many alterations of quantity, physical properties, and chemical composition; so that in judging of a given specimen it is necessary to pursue a methodical enquiry.

The following table copied from Salkowski and Leube gives a summary of the substances met with in normal urine:—

I. ORGANIC.—(1,) Substances belonging to the fatty series:—Urea; uric acid; xanthein bodies; kreatin and kreatinin; oxalic acid; oxaluric acid; volatile fatty acids; glycerin-phosphoric acid; sulphocyanic acid; lactic acid.

(2,) Substances belonging to the aromatic series:— Hippuric acid and benzoic acid; sulpho-carbolic acid; cresolsulphuric acid; sulpho-pyrocatechuic acid; para-oxyphenylacetic acid; parahydrocumaric acid; sulphindigotic acid; skatoxyl-sulphuric acid; cyanuric acid.

(3,) Substances which apparently belong to neither:— Urobilin; sulphur compounds; pepsin; left rotatory substances; cryptophanic acid; extractives.

II. INORGANIC.—Sulphuric acid; hydrochloric acid; phosphoric acid; sodium; potassium; ammonium;

magnesium; calcium; iron; nitric acid; nitrous acid; peroxide of hydrogen; gases.

PHYSICAL PROPERTIES. *Quantity.*—The quantity of urine in health is usually from 40 to 50 ounces, but it may be occasionally as low as 25 or as high as 70 or 80 without indicating disease. In females it is probably less than in males. Yvon and Berlioz give as the mean of their observations 45 oz. (1360 ccs.) in males, and 36 oz. (1100 ccs.) in females, a proportion of five to four. Under pathological conditions the daily quantity may rise to 40 or 50 pints (*polyuria*); on the other hand it may be reduced to a few ounces (*oliguria*), or completely suppressed (*anuria*). In all diseases of the heart or kidneys it is of especial importance to have the urine measured daily; and in hospital practice it should be the rule to have the urine of all patients measured on admission, and from time to time during their stay. Urine-measures of white earthenware can now be procured which have a vertical scale on the inside. The whole of the urine for 24 hours is collected in one of these, and the amount read off from the scale. In this country we usually measure by fluid ounces; on the continent they measure more accurately by cubic centimetres, of which about 30 make a fluid ounce.

Our method is not very exact, but it is not necessary to be so when the urine is fairly abundant; if it is scanty it can be measured more accurately in a glass graduated to cubic centimetres.

The daily amount of urine should find a place on the temperature chart, with the pulse, respirations, and movements of the bowels.

Odour.—The odour of normal urine is peculiar; it has been described as "fragrant" and "aromatic;" but no attempt will be made here to add to these definitions. Certain articles of diet, *e.g.*, garlic and asparagus, communicate to it a very disagreeable smell of sulphides.

Turpentine gives it the odour of violets, not only when taken internally, but when the vapour is inhaled as in polishing furniture or as a medication for bronchitis. Copaiba and cubebs cause the urine to have a peculiar and quite characteristic odour.

Fermentative decomposition inside or outside the body is usually alkaline; by the decomposition of urea ammonia is set free which has its own well-known smell. In diabetes the urine sometimes has a sweetish smell like hay. Urine containing decomposing blood or pus may have a positively putrid odour like decayed fish or flesh.

Translucency.—Normal urine is clear and bright when freshly passed, but may become turbid from precipitation of *urates*, which are insoluble in excess, in the cold. On standing, it may become turbid from decomposition, permitting phosphatic precipitation, or the growth of microorganisms.

Urine may be turbid when passed from the presence of phosphates; such urine is alkaline, and if it is persistent the condition calls for treatment.

Other causes of turbidity are the admixture of mucus, pus, or blood, decomposition in the bladder, and the presence of fat (*chyluria*).

Colour.—The colour of normal urine may vary from the reddish yellow urine of digestion (*urina cibi*), to the nearly colourless urine which follows free potations (*urina potus*).

The normal urinary pigment is *urobilin*, but it is not the sole pigment present, as the spectrum of urine does not coincide with that of urobilin.

Urobilin is formed indirectly from blood-colouring matter (hæmatin). The hepatic cells convert the hæmoglobin first into hæmatin and then into bilirubin; bilirubin is oxidised into urobilin and excreted by the kidney.

Febrile urobilin is an imperfectly oxidised form met with in fevers and in cirrhosis of the liver.

Variations in the yellow colour of urine are for the most part dependent on the proportion of urobilin present.

Certain drugs taken internally give rise to alterations of colour; *e.g.*, *rhubarb*, deep yellow; *santonin*, golden yellow; *chrysophanic acid*, orange yellow; *senna*, brownish; *logwood* and *fuchsin*, reddish; *carbolic acid, tar,* and *creasote*, brown or black. Carbolic acid dressings produce the same effect if the carbolic acid is in excess.

Vogel describes black discolouration of the urine after poisoning by *arseniuretted hydrogen*.

The urine may be coloured *red* by the admixture of *blood*, the tint being deep in proportion to the quantity present; or *brownish*, ranging from smoky to porter coloured, from the formation of methæmoglobin, if the urine has had time to act upon the blood.

In jaundice, *biliary colouring matter* (bilirubin) is excreted by the kidneys, giving rise to various tints of *saffron yellow, mahogany brown, or dark olive green.* When jaundiced urine has been kept some days it may change to a *grass green* colour (biliverdin), from oxidation of the biliary pigment (ROBERTS).

Pus and *fat* (chyluria) give the urine a cream colour, while affecting its translucency. In diabetes the urine is of a *pale greenish* colour. In melanæmia the urine is often *dark brown*. A *blue* colour on the surface is due to crystals of indican.

Darkening of the urine has also been said to be due to alkapton, pyrocatechin, protocatechuic acid, uroleucic, and uroxanthic acids.

Density.—The *density* or *specific gravity* of normal urine is usually from 1·015 to 1·025, but the clear limpid *urina potus* may be as low as 1·005 or less in healthy persons, while urine concentrated owing to free action of the skin in warm weather may be 1·028 or even 1·030.

The normal amount of solids in the urine is about 4 per cent., of which the chief components are urea and common salt; but the proportion of solids to water varies greatly in health, and still more in disease. In a case of post-scarlatinal nephritis W. G. Smith has recorded a urinary density of 1·065; in diabetes mellitus it is generally over 1·030, often over 1·040; in chronic Bright's disease it is usually under 1·015, while in contracting kidney and diabetes insipidus it is very little over 1·000.

The *density* is estimated by means of a *urinometer*, an instrument too familiar to need special description. It is floated in the urine and the density is read off upon a vertical scale. These instruments are not very accurately made; it should repay some instrument maker to give special attention to their construction.

The urine to be examined should be a sample of the whole *mixed urine* for twenty-four hours, a point very commonly neglected and not always attainable; but it may be remembered that great variations occur in the urine in different periods of the twenty-four hours, so that too much importance must not be made of the result obtained from the examination of a stray sample.

If there is too little urine to float the urinometer, the examination may be postponed till it is completed on other points, and then one or more equal quantities of distilled water added till the instrument floats; the last two figures of the result must be multiplied by the number of dilutions to give the true density.

The following precautions in the use of the urinometer should be noted :—

(1,) The glass must not be too narrow. There should be at least half an inch between the stem of the instrument and the side. The cylindrical glasses supplied with urinometers are frequently too narrow.

(2,) The instrument must float freely.

(3,) The surface of the urine should be free from froth; bubbles may be removed by blotting paper.

(4,) The instrument should be free from grease.

Reaction.—Healthy urine is generally *acid*. This is due to the presence of acid phosphate of soda. When this salt is submitted to dialysis, a larger amount of phosphoric acid is found on the outside than on the inside of the dialyser, showing that the acid diffuses faster than the base; this experiment may serve to explain how it is we get an acid secretion like urine from alkaline blood. When the urine is *alkaline* this salt is replaced by alkaline phosphate or by ammonia.

The degree of acidity varies, being diminished partly by food, even so as to render it actually alkaline.

This maximum effect is reached two or three hours after a meal; but, as Roberts points out, no alkaline urine may be voided, because it is mixed in the bladder with acid urine secreted before the meal, or remains there until acid urine is again secreted in a quantity sufficient to modify the reaction.

This alkalinity is due to the presence of fixed alkalies and alkaline earths; on the other hand, the urine passed early in the morning, when many hours have elapsed since the last meal, is excessively acid, but prolonged fasting does not intensify this.

Vegetable and mineral acids increase the acidity of the urine. In phosphaturia, where amorphous earthy phosphates are passed in alkaline urine, this condition can be speedily remedied by the use of nitro-hydrochloric acid; but where the alkalinity is due to fermentative decomposition in the bladder, acids administered internally fail to correct it, probably because they can be given in such small quantities only. The urine is highly acid in gout and allied digestive disorders, in diabetes, acute rheumatism, and chronic Bright's disease.

Alkaline substances readily make the urine alkaline, but the quantities must be large. Roberts says it requires three to four hundred grains of bicarbonate, acetate or citrate of potash, given in divided doses during twenty-four hours, to keep the urine of an adult steadily alkaline.

Lithia is the most powerful base for producing alkalinity of the urine.

Prolonged immersion of the body in a cold bath is said to render the urine alkaline. The urine is frequently alkaline in debility and debilitating diseases, and in the peculiar form of atonic dyspepsia of nervous origin, in which earthy amorphous phosphates are excreted in large quantities.

In all the above cases the urine is alkaline from salts of potash, soda, and the alkaline earths.

Ammoniacal urine is due to decomposition of urea under the influence of a special ferment; this takes place either in the bladder or in the vessel in which the urine is kept after it is voided. Where there is no suspicion of bladder disease the latter possibility should always be excluded by procuring a freshly passed specimen of the urine.

The reaction of urine is estimated by *litmus* paper. This is generally sold in two colours, red and blue, but the latter is alone required, as it soon turns more or less red-violet, and is then most delicate. *Acids* turn litmus paper *red*, while *alkalies* turn it *blue;* the red-violet paper indicates both these changes better than quite blue or quite red paper can do.

Yellow turmeric paper is turned *brown* by alkalies, but it has no special advantage.

Quantitative estimation of the acidity can be performed by means of a standard solution of caustic soda (1 to 10 of water) which is placed in a burette and allowed to run into a beaker containing 100 ccm. of urine, the reaction being tested from time to time by litmus paper till the urine is neutralised. Multiply the number of milligrammes of fluid employed by 0·0063, the quantity of oxalic acid required to neutralise 1 ccm. of the standard solution, and the result gives the percentage acidity (calculated as oxalic acid). For comparative purposes this is quite sufficient.

CHEMICAL COMPOSITION. *Chlorides.*—Common salt or sodium chloride (Na Cl) forms about one fourth of the total solids of healthy urine, and gives the salt taste to this secretion. It is derived from the blood serum, in which it constitutes about 4 parts per 1000, and ultimately of course from the food. It is increased in *ague*, in *diabetes insipidus* and in *Bright's disease*. In Bright's disease complicating pneumonia this increase is not observed (NAUWERCK). It is diminished in *pneumonia*, and in all *febrile* diseases, especially if accompanied by an *exudation;* also in *chorea* and *pemphigus*. In pneumonia a vicarious increase of chlorides is said to be found in the sputa (THUDICHUM).

Chlorides are roughly estimated by precipitating the urine with a ten per cent. solution of *nitrate of silver*, the resulting chloride of silver occupying in healthy urine nearly the whole column of fluid. In pneumonia there is scarcely any precipitate at all.

The quantity of chlorides is not of much importance as a clinical sign, and this method is sufficient for ordinary purposes. If more accurate estimation is required the operation must be performed with a standard solution of silver. The urine having been previously acidulated with nitric acid is repeatedly precipitated and filtered till no further precipitate forms. In accurate experiments the *ash* from a given quantity of urine should be employed.

The quantity of chlorides excreted daily is about ten to twelve grammes.

From thirty to ninety grains of *phosphoric acid* are daily eliminated by the kidneys. It is combined with potash.

Phosphates of soda, lime and magnesia. *Acid phosphate of soda* ($NaH_2PO_4 + H_2O$) gives, as already stated, the acid reaction to urine; a potash salt is also present in small quantity; they are both soluble salts. Phosphoric acid is also combined with *lime* and *magnesia*, forming salts which are soluble in acid urine, but precipitate out when it undergoes alkaline fermentation.

When urine is boiled a milky cloud often forms which is dissolved on cooling or on the addition of a drop or two of acid. This cloud is due to phosphates, and the change that takes place is explained thus by Walter G. Smith:—

$$2\ (\underset{\text{soluble.}}{Ca_2H_2\,P_2O_8}) + \underset{\text{soluble.}}{CaH_4\,P_2O_8} = \underset{\text{insoluble.}}{Ca_3\,P_2O_8} + 2\ (\underset{\text{soluble.}}{CaH_4\,P_2O_8}).$$

On cooling, this process would be inverted.

Salkowski has pointed out that this chemical change may be imitated by a carefully prepared solution of acid calcium phosphate treated with ammonia till a precipitate forms, to which a few drops of acid phosphate are added and the liquid filtered; this solution gives a cloud on boiling which redissolves on cooling or acidulation.

This phosphatic cloud is generally met with in the urine of patients whose digestive powers are feeble or overtaxed, but it is not of serious importance.

A peculiar greenish phosphatic deposit is described by Ehrlich in the urine of typhus, typhoid and measles, when treated with an acid solution of sulphanilinic acid, alkalinised with ammonia, and allowed to stand for twenty-four hours. He regards it as characteristic of these diseases.

When a solution of ferric chlorides is added to urine

a brownish precipitate frequently forms, soluble in excess, due to *phosphate of iron*. This reaction is employed for estimating the phosphoric acid in urine.

An increase of the total phosphates has been observed in acute inflammatory diseases of nerve structures, and temporarily in acute febrile diseases (VOGEL, TEISSIER), occasionally in acute mania and brain tumours, also in chorea, acute yellow atrophy of the liver (BOUCHARD), diabetes (LECORCHÉ), phthisis, chronic rheumatism, leucocythæmia and osteomalacia, and as a primary condition in certain cases of so-called phosphatic diabetes (RALFE, TEISSIER).

A diminution occurs in chronic brain disease, chronic disease of the heart or kidneys, in chlorosis, ague (GEE), rickets, and gout.

The excretion of phosphoric acid is increased by lactic acid and carbonate of soda, diminished by morphia, chloral, ether, chloroform, bromide of potassium (SCHULTZE) and alcohol.

Phosphates of the alkalies do not form urinary deposits, but earthy phosphates are met with as deposits in three forms: (1,) Ammonium magnesium phosphate, or triple phosphate; (2,) Crystalline phosphate of lime; (3,) amorphous phosphate of lime.

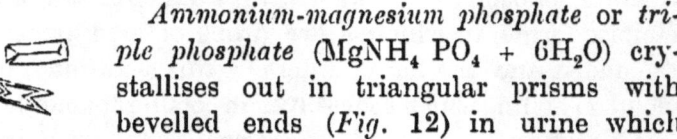

Fig. 12. Crystals of triple phosphate. One prism is incomplete.

Ammonium-magnesium phosphate or *triple phosphate* ($MgNH_4 PO_4 + 6H_2O$) crystallises out in triangular prisms with bevelled ends (*Fig.* 12) in urine which has undergone ammoniacal decomposition, forming a *white* deposit at the bottom of the glass. It may be met with in examining the mucous deposit of a urine which gives an acid reaction, decomposition commencing around the mucus. As a rule this deposit signifies only that the urine has decomposed in the vessel after leaving the body, but it may be met with in fresh urine which has undergone fermentation in the bladder, in cystitis, calculus, myelitis, &c. It is the special constituent of secondary calculous formations, on a nucleus of uric or oxalic acid; or as an incrustation on the walls of a diseased bladder, &c.

Fig. 13. Crystalline phosphate of lime, showing cross, rosette, rod and feather.

Crystalline phosphate of lime. *Acid phosphate of lime* ($CaHPO_4$) occurs in feebly acid urine. It is met with in phthisis, cancer of the stomach and rheumatism, but it may be present in the urine of healthy persons. Its crystals take the form of rods, stars, rosettes, crosses and feathers (*Fig.* 13), hence the names of *stellar* or feathery *phosphate*. It is a somewhat rare deposit.

Neutral phosphate of lime ($Ca_3(PO_4)_2$) occurs in neutral or alkaline urine as an iridescent film on the surface. It has no clinical significance.

Amorphous phosphate of lime ($3CaO\ PO_5$) commonly occurs in urine which is alkaline when passed. It makes the urine turbid and deposits as a copious white sediment, readily soluble in acids. Under the microscope it consists of irregular granules and lumps (*Fig.* 14); on standing, crystals of triple phosphate soon form in it.

Fig. 14. Amorphous phosphate of lime.

It is met with in healthy urine rendered alkaline by food, and in persons who are taking alkalies. As a persistent occurrence it is specially associated with a form of atonic dyspepsia, first described by Prout, which is greatly benefitted by a course of hydrochloric acid or nitro-hydrochloric acid taken regularly after meals.

Fig. 15. Crystalline carbonate of lime (after ROBERTS), spherules and dumbells formed of radiating rodlets.

Carbonate of lime ($CaCO_3$) occurs as an amorphous deposit with earthy phosphates; its crystalline form is very rare. They are spherical bodies composed of numerous radiating needles (*Fig.* 15).

Sulphuric Acid occurs in very small quantity in the urine combined with alkalies, as *Sulphates*, and with indol, skatol and pyrocatechin as aromatic ether-sulphuric compounds.

Sulphuric acid is chiefly derived from the decomposition of proteids, hence its amount runs parallel with the amount of urea excreted (LANDOIS).

The test for it is a solution of *barium chloride*, which gives a copious precipitate of barium sulphate, insoluble in nitric acid.

Sulphur occurs also in cystin, in the sulpho-cyanogen compounds, etc.

Urea ($CH_4 N_2 O$). This substance is present in the urine in a larger quantity than any other solid. Its formation is a function of the liver, and its principal source is the nitrogenous matter taken as food, but it is also formed from the destruction of red blood corpuscles (in the liver) and other tissues, muscle albumen, &c.

The formation of urea is increased by a close atmosphere, such as a kitchen (COOK), by pepsin, maltin, common salt, phosphorus poisoning, arsenic, sulphuric acid, chlorate of potash, hot baths, by excess of nitrogenous food, by coffee, by drugs which stimulate the functions of the liver, *e.g.*, euonymin, corrosive sublimate, salicylic acid, benzoic acid, colchicum, by fever, by active congestion of the liver without destruction of its substance, by pernicious anæmia, malaria, &c. It is not increased by muscular exercise.

Its elimination is proportional *(cæteris paribus)* to the amount of urine excreted, so that copious drinking of water or any fluid, increases the elimination of urea; it is also increased in diabetes mellitus and insipidus. Willis described and Prout believed in a pathological condition characterized by excess of urea excretion, to which the name *azoturia* was given, but there is no such disease as an independent condition.

The formation of urea is diminished by fasting, by drugs which depress the function of the liver, *e.g.* lead, and by those diseases of the liver which depress its function or destroy its substance.

Its elimination is checked by anything which diminishes the amount of urine, by profuse sweating, by diseases of the kidneys and urinary apparatus, by wasting diseases, acute gout, chronic rheumatism, lepra, pemphigus, melancholia, imbecility, catalepsy, hysteria and cholera. It is supposed by some that the excretion of urea is diminished in a very early stage of contracting Bright's disease, before structural alterations have occurred in the kidney; but this so-called *renal inadequacy* is quite fanciful.

The daily quantity of urea varies at different ages; according to Ralfe

At 5 years	180	grains
12 ,,	320	,,
21 ,,	535	,,
40 ,,	555	,,

Camerer gives the proportion in children at 0·64 to 1·12 grammes per kilogramme of body weight, while in adults the proportion is only 0·5 to 0·6 grammes per kilogramme.

Women excrete absolutely less than men, but not relatively in proportion to their weight.

The estimates for the inhabitants of different countries shew variations, which are probably accounted for by differences in diet:

Englishmen (RALFE) -	- 555	grains
Bavarians (VOIT, RUBNER) -	425	,,
Frenchmen (YVON, BERLIOZ) -	397	,,
North Germans (FLUGGE) -	352	,,

One day's fasting is enough to reduce the urea excretion

to two hundred and eighty-eight grains (RANKE), while after several days it falls to ninety grains (SCHULTZEN).

In acute Bright's disease, when the urine is greatly reduced, the elimination of urea is necessarily also diminished; but the percentage of urea which should be high if there were merely a reduction of the urinary water, is *low*, generally below two per cent.

In sub-acute Bright's disease, as the urine increases in quantity the percentage of urea tends to fall lower, averaging about one per cent.

In the later stages the percentage of urea remains low while its total amount varies with the quantity of urine; but as this is sometimes high it may reach a fair figure. On light diet, with chicken and fish, such patients may excrete from three hundred to four hundred grains of urea.

In the lithæmic form the urine throughout the early and middle course of the disease is much increased in amount, except during the occurrence of an intercurrent attack of acute nephritis, to which these cases are no doubt very liable. In these early stages and even later the amount of urea may be normal.

W.S., aged twenty-three, was sent to me from the Eye Hospital with bilateral diffuse neuro-retinitis. (Case 4, p. 83). He was admitted on the 24th February, 1888, and on the 26th his urine was analysed; it amounted to seventy-eight ozs. in twenty-four hours, and contained 0·8 per cent. of urea, equivalent to two hundred and ninety grains in twenty-four hours. On chicken diet he passed three hundred and seventy grains. He died in uræmic coma on April 7th, and his kidneys were found to be in a condition of advanced contraction. This case supports the statements of Bartels, Grainger Stewart and others, that so long as

Fig. 16. Crystals of nitrate of urea artificially prepared by the addition of nitric acid to urine.

there is polyuria, the elimination of urea may not be diminished.

Detection of urea. The presence of urea may best be demonstrated by adding strong nitric acid to a little of the concentrated urine or other fluid in a watch glass, when crystals of nitrate of urea will form, which can be recognised by the microscope (*Fig.* 16); any albumen must be first removed by boiling and filtration.

Estimation of urea.—The most convenient method of estimating urea is by decomposing it with *hypochlorite or hypobromite of sodium* solution, the amount of urea being determined by measuring the volume of nitrogen evolved. This method is sufficiently accurate for clinical purposes. The cheapest and simplest form of apparatus is that invented by Professor Doremus, of New York, and sold by Messrs. Southall, Birmingham. (*Fig.* 17).

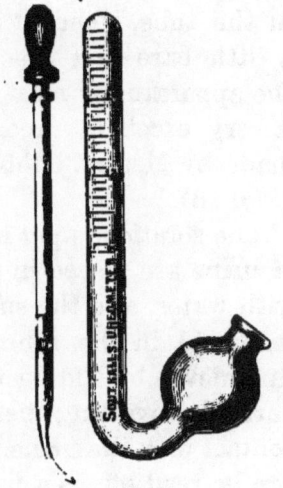

Fig. 17.

The solution is poured into the tube till it half fills the bulb, then some urine is drawn up into the pipette as high as the scratch, the beak of

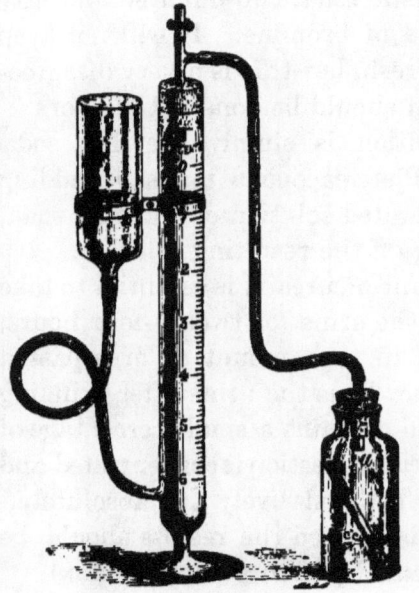

Fig. 18.

the pipette is carefully inserted into the curve of the tube, and the urine is made to flow by gentle pressure into the solution, the gas being given off and collected in the tube, where it is read off as a percentage. With a little care and practice there is very little loss, and the apparatus is most manageable and very inexpensive. A very excellent ureometer is that known as Gerrard's, made by Messrs. Gibbs, Cuxson and Co., of Wednesbury. (*Fig.* 18).

The solution is put into the little tube, and twenty ccm. of urine are placed in the bottle. The apparatus is filled with water, and the surfaces of the fluid in the graduated tube and in the funnel at the side are brought to the same level by sliding the funnel up or down. Then by partially inverting the bottle the solution is brought into contact with the urine, the gas is liberated, and the result can be read off as a percentage in the graduated tube.

The hypobromite solution is more energetic and rapid in its action than the hypochlorite. Its composition is one hundred parts of caustic soda, two hundred and fifty of water, and twenty-five of bromine. It will not keep well, and is better made fresh, but this is a very disagreeable smelling process, and should be done out of doors.

The hypochlorite solution is simply the liq. sodæ chlorinatæ of the U.S. Pharmacopœia made by adding chloride of lime to a saturated solution of washing soda, and filtering or siphoning off the resulting solution.

In estimating the amount of urea it is essential to take a sample of the whole of the urine for twenty-four hours, as great variations occur in the amount of urea passed under different conditions. Thus the urine after drinking fluids freely is copious and contains a small percentage of urea, while that passed during digestion is concentrated and contains a great quantity both relatively and absolutely.

When great accuracy is desired the results should be corrected for temperature and pressure (NOEL PATON).

CLINICAL EXAMINATION OF THE URINE. 113

Uric acid ($C_5H_4N_4O_3$) is normally present in urine as urate of soda, which is not decomposable by the acid phosphate of soda to which the urine owes its acid reaction. If any free acid is liberated the urates are decomposed, and uric acid crystals are thrown down.

It is originally formed by the splitting up of albumen, but how and where this process takes place is doubtful, though a good deal of evidence points to the spleen as the seat of the change.

The normal daily amount excreted is seven to ten grains; it is not increased by nitrogenous diet (GARROD). Cook found beef tea, Liebig's extract, pepsin, maltin, euonymin, and confinement in a close atmosphere increase the amount of uric acid.

It may be sometimes increased by the excessive use of *milk*.

The urine of newly born children contains much uric acid; it is increased in febrile diseases, acute rheumatism, pneumonia, ague and malarial fevers, leucocythæmia, cirrhosis of the liver, and diabetes.

It accumulates in the blood in gout (GARROD), and is excreted in greater quantity after an attack.

Its excretion is increased by colchicum, corrosive sublimate and euonymin, diminished by salicylate of soda and benzoate of soda (NOEL PATON). It is also diminished in anæmia, chlorosis and gout, and by copious draughts of water, large doses of quinine, caffein, iodide of potassium, common salt, carbonate of soda and lithia, sulphate of soda, oxygen inhalation and slight muscular exertion (LANDOIS AND STIRLING).

Fig. 19. Crystals of uric acid in various forms.

Crystals of *uric acid* form a deposit like cayenne pepper on the bottom or sides of the glass; if larger, they may look like crystals of brown sugar. They assume

8

various forms,—lozenges, rhomboids, hexagons, stars, spikes, &c. (*Fig.* 19).

The *murexide* reaction is the ordinary chemical test; a little of the suspected matter is heated slowly with a drop of nitric acid on a porcelain dish, and allowed to cool; a drop of dilute ammonia is then added, when a purple red colour due to *murexide* is developed if uric acid is present.

Quantitative estimation may be performed by adding five ccm. of concentrated *hydrochloric acid* to a hundred ccm. of urine and allowing it to stand for forty-eight hours in the dark, when the uric acid deposits and may be dried and weighed. Haycraft has devised a much better method by precipitating the uric acid as a silver salt.

Amorphous urate of soda is normally present in urine, forming a reddish deposit in concentrated urine after it has cooled; it is also a common deposit in febrile diseases, and in cirrhosis and other affections of the liver. Under the microscope it appears as small irregular granules (*Fig.* 20), while on the addition of acetic acid uric acid crystallises out. According to Roberts it consists mainly of quadrurates which are readily decomposed by water, but this change is prevented by the salts of the urine, especially by the sodium chloride and potassium phosphate. Crystalline urates are rare.

Fig. 20. Amorphous urate of soda.

Fig. 21. Hedgehog crystals of urate of soda.

Crystalline urate of soda is sometimes deposited in the urine of children during febrile attacks. It forms characteristic yellow hedgehog crystals (*Fig.* 21). Roberts attributes the frequency of vesical calculi in children to this deposit.

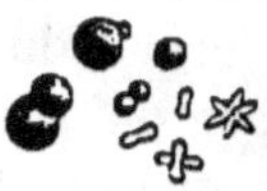

Fig. 22. Urate of ammonia in dark yellow spheres, pale dumbells, crosses and rosettes.

Urate of ammonia is met with in ammoniacal urine; it has no clinical significance. It forms dark yellow spheres and pale slender dumbells (*Fig.* 22).

Hippuric Acid ($C_9H_9NO_3$) occurs to a small extent in human urine; it is formed when *benzoic* acid or some nearly related chemical body is introduced into the alimentary canal; and is also formed in the body from proteids (LANDOIS and STIRLING).

Fig. 23. Crystals of hippuric acid from the urine of the horse.

It crystallises in colourless four-sided prisms (*Fig.* 23.)

In the herbivora it appears to replace uric acid to a great extent in their normal urine.

Kreatinin ($C_4H_7N_3O$) is derived from muscle kreatin; it is a normal constituent of urine. It may be detected by adding to the urine a few drops of a "slightly brownish" solution of nitro-prusside of sodium and weak caustic soda, when a burgundy red colour is developed which soon fades. On heating with acetic acid the colour changes to green or blue.

It is increased in fevers, pneumonia, etc., diminished in anæmia and wasting diseases; it is not diminished by fasting.

Xanthin ($C_5H_4N_4O_2$) also occurs in normal urine in very small quantity. When evaporated with *nitric acid* it gives a yellow stain which becomes yellowish red on adding *potash*, and red violet on heating.

It sometimes forms calculi, which are of a deep yellow colour, smooth and spherical.

Sarcin or Hypoxanthin ($C_5H_4N_4O$) has been found only in the urine of leucocythæmia, though a body nearly related to it occurs in normal urine (SALKOWSKI). When evaporated with *nitric acid* it gives a light *yellow* stain, which becomes deeper but not reddish on adding caustic soda.

Succinic Acid ($C_4H_6O_4$) occurs chiefly after a diet of flesh and fat, after eating asparagus and after drinking alcohol.

Lactic Acid ($C_3H_6O_3$) is present in normal urine. Colasanti and Moscatelli found it present in large quantities in the urine of soldiers after a march, and they suggest that it is probably a product of muscular activity which passes into the urine.

Oxaluric Acid ($C_3H_4N_2O_4$) is present in traces in normal urine combined with ammonia. It is a derivative of uric acid, and on being heated splits up into urea and oxalic acid.

Oxalic acid ($C_2H_2O_4$) is a normal constituent of urine, and in combination with soda and potash remains in solution. It is probably formed by the oxidation of uric acid into oxaluric acid ($C_3H_4N_2O_4$), which splits up into oxalic acid and urea.

Oxalate of Lime ($C_2CaO_4 + 2\,H_2O$). Crystals of oxalate of lime can always be found among amorphous urates, if the urine has stood some time, but they may be deposited alone or with a little mucus in a very characteristic fleecy cloud; when not very abundant they may be seen glittering on what at first sight looks like a simple mucous cloud.

Oxalate of lime is formed by the decomposition of oxalate of soda and potash, and the union of the acid with lime salts.

A fleecy deposit of oxalate of lime crystals is associated with a special form of atonic dyspepsia to which the name of *oxaluria* has been given.

The precise conditions under which this deposit occurs cannot be formulated, but debility, anæmia, and various organic diseases of all kinds are predisposing causes. Over-eating under certain conditions may be a direct cause.

Fig. 24. Crystals of oxalate of lime, octohedra, pyramids, dumbell and spheroid.

The crystals can readily be recognised under the microscope with a power of two hundred and fifty diameters by their shape and high refractive powers. They are generally octohedral or pyramidal, more rarely dumbell shaped and not uncommonly spherical or oval (*Fig.* 24). These spheres are said by Roberts to be dumbells seen endwise, and Beale speaks of them as discs. I have frequently failed to see a single dumbell where these spheroids were visible, or to transform them into dumbells by shifting them about.

Cystin ($C_3NH_7SO_2$) is a left rotatory body which occurs normally in the urine in very small quantities, but very rarely in so large amount as to give rise to a deposit or the formation of calculi. Its formation is altogether obscure; the tendency to overproduction being probably congenital and hereditary. It is recognised by its characteristic crystals in the form of hexagonal plates (*Fig.* 25).

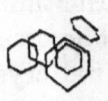

Fig. 25. Hexagonal plates of cystin.

Cystin is insoluble in water, ether, and alcohol, but very soluble in ammonia and caustic alkalies.

A very good test for cystin is to boil some of its potash solution, dilute it with water, and add a little nitro-prusside of potash, when a beautiful violet colour is developed.

Leucin ($C_6H_{13}NO_2$) and *Tyrosin* ($C_9H_{11}NO_3$) occur in the urine in acute yellow atrophy of the liver. They are products of the pancreatic digestion, but are normally further oxidised into urea.

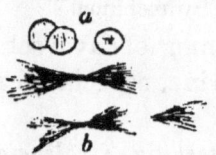

Fig. 26. *a*, Balls of leucin which with polarised light shew a dark cross. *b*, Sheaves of tyrosin.

Leucin, precipitated spontaneously or obtained by evaporating an alcoholic extract of the urine, occurs in the form of yellowish brown balls. Tyrosin forms silky colourless sheaves (*Fig.* 26) of needles. When slightly heated with a few drops of concentrated sulphuric acid it dissolves with a temporary deep red colour.

According to Anderson both are frequently present in the urine of liver diseases.

Indican ($C_8H_6NKSO_4$) is formed in the intestine by the pancreatic digestion or by putrefactive change. It is derived from indol (C_8H_7N). By oxidation it forms indigo-blue.

It is increased in the urine in intestinal diseases, prolonged constipation, hernia, typhoid fever, cancer of the bowel, in Addison's disease, diabetes, Bright's disease, nervous diseases, peritonitis, cholera, fractures, osteomyelitis, etc. The largest quantity I have ever seen was in the very scanty urine of a hysterical patient. Ord has described a calculus formed from it.

Indican is the substance formerly called uroxanthin; it gives the urine a deep yellow colour; but if it becomes oxidised in decomposing urine a film of bluish red crystals of indigo-blue is formed.

The best test is to add to equal quantities of concentrated hydrochloric acid and urine, drop by drop, a solution of hypochlorite of lime and shake until a blue colour is developed; if the mixture is shaken up with chloroform the blue colouring matter is taken up by it and can be obtained as a deposit when the chloroform evaporates.

Hydrochinon ($C_6H_6O_2$) is the substance which gives the dark colour to urine in cases of carbolic acid poisoning. It is formed by oxidation.

$$\underset{\text{carbolic acid.}}{C_6H_6O} + \underset{\text{oxygen.}}{O} = \underset{\text{hydrochinon.}}{C_6H_6O_2}$$

Pyrocatechin ($C_6H_6O_2$) is an isomer of hydrochinon, and also gives a dark colour to urine, especially where putrefaction has occurred.

Alcapton; Pyrocatechin; Protocatechuic, Uroleucic and Uroxanthic Acids.

Bödeker discovered a substance in the urine which darkens in the presence of alkalies, named by him alcapton.

This is said by Ebstein and Müller to be pyrocatechin, while Walter G. Smith thinks it is protocatechuic acid.

Kirk believes it to be a compound body containing at least two components, which he has named uroleucic and uroxanthic acids.

Grape sugar ($C_6H_{12}O_6$) is present in traces (0·0002 per cent.) in normal urine, but in larger quantities in the urine of diabetes mellitus.

Milk sugar is said to be sometimes present in the urine of pregnant or nursing women.

Physiological traces of sugar are not recognisable by the ordinary tests.

Temporary or intermittent functional glycosuria occurs, but is rare.

In various serious organic diseases of the brain and spinal cord glycosuria may be occasionally met with.

The mode in which sugar is produced is not yet fully understood. Glycosuria occurs when the centre for the hepatic vaso-motor nerves in the floor of the fourth ventricle is punctured, or after section of the vaso-motor channels in the cord, or section of the vaso-motor nerves going to the liver. Glycosuria so induced may be stopped by section of the splanchnic nerves. Poisons which paralyse the vaso-motor nerves of the liver, e.g., chloroform, ether and chloral, produce the same effect.

The prevalent explanation of these facts is that through the agency of a ferment the blood normally converts the glycogen of the liver into sugar in such quantities as can be oxidised in the lungs into water and carbonic acid; but when there is excessive afflux of blood, as in vaso-motor paralysis, the ferment in the blood converts more glycogen into sugar than can be burnt off, and the excess of sugar appears in the urine.

Modern observations make it more probable that the conversion is a function, not of a blood ferment, but of the activity of the liver cells on food principles, sugar, peptone, and fat.

Seegen has shown that the liver has the power of converting peptone and fat into sugar, which explains the persistence of glycosuria when saccharine and starchy food is withheld. Sugar is also produced by catalytic changes in muscle, but it is not probable that enough is ever produced in this way to cause diabetes. Seegen thinks that there may be a condition of cell life throughout the organism in which these elements have lost the capacity for destroying the sugar brought to them by the blood.

The best clinical test for sugar is *Fehling's* solution, prepared according to the following formula:—

Sulphate of Copper	90½ grs.	Solution of Caustic Soda	
Neutral Tartrate of Potash	364 grs.	sp. gr. 1·12	f℥iv

Add water to make up exactly six fluid ounces.

Two hundred grains of this solution are exactly decomposed by one grain of sugar (ROBERTS).

Boil about a drachm of the solution in a test tube, add an equal quantity of urine, boil again; if sugar is present the yellow suboxide of copper is thrown down. Fehling's solution does not keep indefinitely, so that it is necessary to make a blank experiment by boiling the solution before adding the urine, when the cuprous oxide will be thrown down if the Fehling is too old.

Albumen causes a dirty purple precipitate, peptones turn Fehling rose pink, and uric acid or urates cause a slight reduction of copper. In cases of doubt it has been suggested to prepare two test tubes containing urine and Fehling, and adding a little yeast to one, keep them in a warm place all night, then boil them both next morning. If sugar is present the slight reduction will not take place in the tube to which the yeast was added but will appear in the other.

A more ready plan is to employ another test. Put about half a drachm of a saturated solution of *Picric acid* in a test tube, add a few drops of dilute *Liq. potassæ* (1 to 10), boil the mixture, add an equal quantity of urine and boil again; if as much as one grain of sugar per ounce be present the liquid will become quite *opaque*, but a certain amount of reddish brown colouration takes place whether sugar be present or not.

The *quantitative estimation* of sugar is made with Fehling's solution. The necessary apparatus consists of two burettes, a glass flask, or white porcelain dish and a spirit lamp. Measure off two hundred grains of the solution in one of the burettes, run it into the flask or dish, dilute it with two volumes of distilled water and set it on to boil. Dilute the urine with distilled water to ten volumes, fill the other burette with it, let it run drop by drop into the boiling flask or dish, agitating or stirring it the while, and from time to time removing the lamp so as to let the reduced copper settle. When the blue colour has entirely disappeared from the fluid read off the quantity of urine which has been required to effect this; this quantity contains exactly one grain of grape sugar. Let us suppose it was a hundred and fifteen grains of the diluted urine, which represents 11·5 grains of undiluted urine; then by dividing a hundred by this amount we get, namely, $\frac{100}{11\cdot5}$ = 8·69 grains per cent. The method is simple enough and does not take a long time, but yet it is perhaps more than every busy practitioner can always manage to give. It may therefore be stated that if the patient is made to weigh himself and measure his urine from time to time, no quantitative analysis is really necessary to judge of the progress of the case.

The *polarimeter* is sometimes recommended for estimating sugar, but in my hands it has proved very unsatisfactory.

Acetone (CH_3COCH_3) is met with in the urine of *bad* cases of diabetes, of many acute febrile diseases, e.g., measles, scarlatina, pneumonia, after the ingestion of alcohol (BULL), and even in the urine of healthy children (BAGINSKY).

The best test is that devised by Le Nobel. Pour about an ounce of urine into a urine glass, add a drachm or two of a solution of nitro-prusside of sodium (5 grains to 1 oz.), and a few drops of strong *ammonia*. After standing a few minutes a rose violet colour is developed, which, if much acetone is present, will require diluting with water to bring out the brilliancy of its colour.

Acetone is probably formed in the urine by the breaking up of aceto-acetic acid into acetone and carbon dioxide.

$$CH_3,CO,CH_2,COOH = CH_3,CO,CH_3, + CO_2.$$

Aceto-acetic acid or *Diacetic acid* ($CH_3,CO,CH_2,COOH$) is present in the urine of diabetes, and gives a red coloration with *ferric chloride solution*, which disappears on heating. This *ferric chloride reaction* was at one time thought to be a test for acetone. According to Le Nobel the same reaction is given by β-oxybutyric acid, sulpho-cyanogen, acetic and formic acid compounds, only differing with these by not disappearing on the application of heat.

The large number of substances which give this reaction accounts for its being so commonly met with, as in coma, chronic Bright's disease, perityphlitis, strangulated hernia, after minor surgical operations, and in sulphuric acid poisoning (WINDLE). It is not of any clinical significance even in diabetes.

Bile pigment or *Bilirubin* ($C_{32}H_{36}N_4O_6$) is present in the urine in cases of jaundice, sometimes before any change is to be observed in the skin or conjunctivæ.

It colours the urine deep yellow, mahogany brown or olive green. The last colour is due to partial oxidation of the bilirubin to *biliverdin* ($C_{32}H_{36}N_4O_8$). Bile stained urines may turn *grass green* from this change, which only takes place when the urine is undergoing decomposition.

The peculiar colour of bile-stained urine can be usually recognised by the eye, even when only a small quantity of the pigment is present; but some dark urines look very much as if they contained a large quantity of bile pigment until they are well diluted with water.

The best way of testing for bile pigment is to dilute a couple of ounces of the urine to the colour of sherry in a urine glass, add a drachm or two of strong *nitric acid* containing some *nitrous acid* (such nitric acid has a yellow colour), or weak liq. iodi (1 to 10 of water); if bile pigment be present a grass green colour is developed. Sometimes one and sometimes the other of these tests gives the best result.

The method by dilution in a urine glass is recommended as giving much better results than the plan ordinarily followed of making a play of colours by mixing a drop of urine with a drop of nitric acid on a white porcelain dish.

Bile acids, glycocholic ($C_{26}H_{43}NO_6$) and taurocholic ($C_{26}H_{45}NSO_7$) acids are often present in the urine of jaundice.

The following test is recommended by Hay. Sprinkle a little *precipitated sulphur* on the surface of the urine. If bile acids are present the grains of sulphur *float*, instead of sinking as they otherwise do.

Fat or oil globules may be found in the urine after the passage of a catheter, and are of course derived from the oil used on the instrument; they are met with also in phthisis, pyæmia, long standing suppuration, and phosphorus poisoning, from fatty degeneration of pus, or of renal or vesical epithelium; but in all these conditions are present only in quantities recognisable by the microscope.

In *chyluria* the urine is milky from admixture of fat. This condition occurs sometimes in pregnancy and lactation. Mr. Frost, of Yardley, a year or two ago brought me the urine of a young unmarried girl, who, having become pregnant, had compressed her abdomen so much in order to conceal her condition as to cause œdema of the legs, thighs, vulva and lower part of the abdomen. After her confinement the urine became milky and remained so for some days; it contained fatty granules, cholesterin and albumen, but no sugar.

Francotti has described a somewhat similar case in which a woman who had never resided in the tropics passed chylous urine in each pregnancy. With rest it diminished or disappeared, recurring on going about.

In both these cases there was probably some rupture of dilated lymphatics with escape of lymph or chyle into the urine.

Rossbach has published a case of a young girl with mitral insufficiency who passed milky urine both by day and by night. The daily quantity of fat excreted varied from 1·5 to 10 grammes. The urine also contained albumen. The patient had diminished liver dulness, and he suggests that it is possible the condition depended in some way upon disease of the liver.

Endemic chyluria occurs commonly in India, China, and the Straits in persons whose blood is infested by the parasitic nematode, *filaria sanguinis hominis*. Chyluria is caused when the lymphatics get obstructed through the impaction of abortive ova in the ducts. (MANSON).

Fatty acids, *e.g.*, butyric acid, are sometimes present in the urine.

Cholesterin ($C_{24}H_{44}O$) is met with in chylous urine, in fatty degeneration of the kidney, in diabetes, jaundice, and in the urine of epilepsy treated with potassium bromide (PÖHL), while it enters into the constitution of certain urinary calculi.

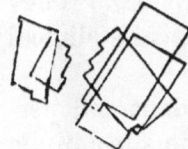

Fig. 27. Crystals or plates of cholesterin.

Its crystals can be recognised under the microscope as large thin rhomboidal colourless plates (*Fig.* 27). With iodine and sulphuric acid they turn a deep blue or violet colour.

A further test is made by dissolving the cholesterin with chloroform in a test tube, adding concentrated sulphuric acid and shaking for some minutes; the chloroform in the presence of the slightest trace of cholesterin becomes citron yellow, with a larger quantity, blood red or purple. On the addition of water to the chloroform solution it becomes quickly blue, then green, and finally yellow. With glacial acetic acid the violet solution shows a green fluorescence.

Albumen. By albumen is meant serum albumen, with its acid and alkaline modifications, and serum globulin (paraglobulin).

It must be borne in mind that the clinical examination of the urine is not a scientific enquiry, but a practical method founded upon empirical as well as upon scientific data. There are albuminoid bodies in the urine which have no known clinical significance, and these can only embarrass the practitioner who finds them when he is looking for a sign to which he attributes a significance based upon a purely empirical foundation. Posner has proved that a minute amount of serum albumen is a constant constituent of normal urine, so that if we could obtain a very delicate test for albumen it would be actually misleading to us.

I have defined albuminuria as *the presence in the urine of a substance which is coagulable by heat or precipi-*

tated by neutralisation. This definition includes serum albumen, with its modifications, and serum globulin, while it excludes a variety of albuminous substances which are sometimes present in urine, but concerning whose pathological relationships we know very little.

The conditions under which albuminuria occurs have been considered in a previous chapter (Chap. I.). They indicate that it is very wide-spread, and that its clinical significance is proportionately vague.

In Bright's disease albuminuria is abundant and persistent in acute and sub-acute attacks, and in the chronic affection following such attacks; but in the latent form of Bright's disease, in contracting kidney in its most typical form, the amount of albumen is small, and it may be altogether absent, though this is rare—rarer in my experience than has been thought by some writers.

In order to find minute traces of albumen the following method and precautions should be carefully adhered to. Albumen is most constantly present in the urine passed in the forenoon, so that it is best to obtain a specimen passed at that time if possible.

Putrid urine is unfit for detecting traces of albumen, and turbid urine, unless the turbidity be due to urates which will disappear on heating, should be filtered.

Fill a test tube two thirds full of urine, hold it by the lower end and boil the upper half *well* over a spirit lamp, then acidulate with two or three drops of *dilute acetic acid*. If there is any difficulty in recognising a cloud, hold the tube against a shaded back-ground with the light falling from above, when the faintest haze will be apparent by contrast with the layer of unboiled clear fluid below. If the room is badly lighted, or the examination is made by artificial light, a very faint haze may escape detection.

This haze indicates with certainty the presence of serum albumen or serum globulin. There is no single operation by which these can be differentiated. If it is desired to do so a portion of the urine must be saturated with common salt or with magnesium sulphate and filtered. By this means the globulin is precipitated, so that if a urine before precipitation gave a cloud on boiling and acidulating, but none afterwards, we should be justified in concluding that the albuminous body was serum globulin. But globulin very rarely occurs by itself, and when it does it has not been proved to have any significance, though Hermann has stated that he found globulin only in a case of eclampsia, while in the nephritis of pregnancy he found serum albumen. It has been repeatedly suggested that in various forms of "functional" albuminuria, the albuminous body is globulin, but this statement has hitherto in all cases been proved to be erroneous; there can be no doubt that the presence of globulin unaccompanied by albumen is very exceptional and has at present no definite significance.

It may be contended that it would be better to have a test which does not throw down globulin, and this may be readily granted, but there is no test which is perfect. This trifling imperfection will never lead any one astray who bears in mind *that the discovery of albuminuria is not a fact "per se" of definite pathognomonic significance.*

It has been objected that urine which contains oxalate of lime gives a reaction resembling albumen with this test. This is true if urine is saturated with oxalic acid, filtered and then tested, but the haze is a *very* faint one, and it is not true in my experience of oxaluria as observed clinically.

One observer has stated that mucin gives a cloud resembling albumen, but this is not the case. On adding acetic acid mucin coagulates in tiny filaments easily distinguished from albumen.

Heller's test, with *cold nitric acid*, is a very good plan. It is performed by pouring a few drops of nitric acid into a test tube and then floating half a drachm of urine by means of a pipette on its surface. Where the liquids come in contact a cloud forms when albumen is present.

The objections to this method are: (1,) That it is not so delicate as the heat and acetic acid plan, though, like that, it throws down globulin; (2,) That the reaction is simulated in the urine of patients taking copaiba, cubebs or salicylic acid; (3,) That nitric acid is hurtful to one's fingers, clothes and carpets, when it gets spilt, as in the course of daily use it is certain to be some time; (4,) That it gives a cloud with uric acid, and that urea may crystallise out.

The use of *picric acid* has been strongly advocated of late years. It is recommended to employ a double saturated solution of common salt and picric acid, which is heavier than most urines, and it is then used like cold nitric acid as just described, a cloud forming where the two liquids come in contact.

The objections are: (1,) That this reaction occurs with *peptone*, an albuminous body of no known pathological significance, which is frequently present in urine; with alkaloids, such as quinine, and with mucin.

The cloud given by peptone and alkaloids disappears on heating; but if the spirit lamp is to be employed in every case the test loses its principal advantage of simplicity. Unfortunately the fallacy from mucin cannot be got rid of except by testing separately for mucin. (2,) Picric acid is not so delicate as either of the above described methods. There is some conflict of testimony on this head, but this is my experience, and it is supported by Roberts, Lauder Brunton, Harris, and Stirling.

Ferrocyanide of potassium and *acetic acid* give unsatisfactory results in the examination of urine. They throw down propeptone, as well as albumen proper.

Tanret's test, *potassio-mercuric iodide*, throws down

CLINICAL EXAMINATION OF THE URINE.

peptone as well as albumen, the cloud produced by the former substance not disappearing on heating.

Tungstate of soda possesses the same fallacious power, and also gives a cloud with mucin.

It is unnecessary to go further through the various tests which have been recommended. It is possible that individuals may find themselves able to get better results with one test than another; but the three methods described fully, namely, boiling and acetic acid, cold nitric acid and picric acid, are those which have received the support of the best authorities; all others are for various reasons useless or fallacious.

When a *quantitative analysis* is required the most accurate way is to boil a known quantity of urine, acidulate it, and collect it on a weighed filter paper, dry it over sulphuric acid in a bell jar, and weigh it carefully, deducting the weight of the paper. But this is a method unsuited to the needs of practitioners.

Fig. 28.

A very easy and fairly accurate method has been invented by Esbach. The only apparatus required is a specially graduated tube* (Fig. 28). The urine, diluted with one or more volumes of water, is poured into this tube up to the line marked U, and the albumen is precipitated by a solution of *picric* and *citric acids* (ten grammes of picric acid, and twenty grammes of citric acid, dissolved in eight hundred or nine hundred ccm. of boiling water, and enough water added to make up one litre), the tube being then allowed to stand a few hours when the result is read off as a percentage, and corrected according to the number of volumes of water added.

When we wish to estimate the amount of albumen by

---

* These tubes are sold by Messrs. Southall, and Philip Harris, of Birmingham.

either method, it is necessary to make use of a sample of the whole twenty-four hours' urine in order to give the observation any value, as the proportion of albumen varies very much at different times of the day and night.

Peptone has no known pathological significance, but is met with in many acute diseases. The test for it is that it gives a *pink* colour with Fehling's solution in the cold; it is better to dilute the reagent with an equal quantity of distilled water.

Mucin is commonly present in the urine of even healthy persons, causing a light flocculent cloud which gradually settles to the bottom of the vessel. In catarrh of the urinary passages it is much increased and mixed with pus.

It is precipitated by acetic acid in the form of fine filaments, and by alcohol, citric acid, picric acid, tungstate of soda, &c. After boiling with hydrochloric acid mucin reduces cupric oxide like sugar.

Blood.—The presence of blood in the urine is a symptom common to a number of pathological conditions, differing essentially in their seat, nature, and relationships. It may appear in a corpuscular or non-corpuscular form; the latter is called hæmoglobinuria.

The urine of women is bloody during menstruation, and whenever there is vaginal or uterine hæmorrhage. Hæmorrhage from the urethra may be caused by villous growths, or in consequence of local congestion or injury. The blood is bright red, appears independently of micturition, or is not mixed with the stream, but occurs at the beginning or end of it, and is often accompanied by local pain or other symptoms.

Hæmorrhage from the bladder may be caused by stone, prostatic disease, villous or malignant growths, cystitis, ulcer, parasites (*Bilharzia*), etc. In stone, prostatic disease, and cystitis, the diagnosis is not difficult, as these conditions have well marked symptoms. The first two can soon be excluded by physical examination, while parasitic ova may be recognised by the microscope. But ulcers and growths in the bladder present peculiar difficulties, which may long baffle diagnosis. Here the cystoscope promises to be of great assistance.

Hæmorrhage from the bladder is usually associated with some degree of cystitis and local pain, frequency of micturition, etc. By passing a sound or lithotrite, fragments of growth may be obtained or an irregular ulcerated surface detected. Washing out the bladder may afford useful aid in obtaining fragments of villous growth.

In women, urethral dilatation and digital exploration constitute a safe and easy method of examining the inside of the bladder, while in males, after due consideration, an exploratory cystotomy may be performed.

Hæmorrhage from the renal substance usually reveals itself by blood casts of the urinary tubules, but hæmorrhage from the pelvis has no such constant sign, though casts of the ureter may be found. Renal hæmorrhage is usually accompanied by local pain, while the history of injury, a blow, passage of calculus, etc., may help.

Its causes are very numerous :—

1. *Local Lesions.*—External injury, twisted or movable kidney, calculus, tubercle, cancer, syphilis, embolism, parasites, congestion, Bright's disease.

2. *Symptomatic.* — Blood diseases (purpura, scurvy, hæmoglobinæmia, leucocythæmia), specific fevers, malaria, cholera.

3. *Toxic.*—Turpentine, cantharides.

4. *Neurotic or Vicarious.*—Hysteria, insanity, asthma, menstruation, hæmorrhoids.

Detection of Blood in the Urine.—The diagnosis of the presence of blood colouring matter in the urine may be made by (1,) The eye; (2,) The miscroscope; (3,) The guaiacum test; (4,) The spectroscope; but of these the microscope only is capable of differentiating hæmaturia from hæmoglobinuria.

It has been maintained by Wickham Legg that the blood corpuscles are always broken up after the urine is secreted. He maintains that if the urine is examined immediately after leaving the body, corpuscles can always be found.

The following case proves that this is not always the case:—

CASE 7. *Paroxysmal Hæmoglobinuria.*—K. G., aged 28, a labourer, was admitted into hospital on February 22nd, 1886, complaining of shivering, pain in the back and legs, and bloody urine, which had lasted since the previous morning. He had been subject to these attacks for the last three winters; the first came on while working in a cold wind in the early winter of 1881. There was no history of ague or residence in a malarial district. He had had syphilis,—a chancre followed by a skin eruption ten years before. He fell from a scaffold in 1879, and had a blow on the head three years ago; but these accidents were not in any way related to his attacks. While in hospital he had a well marked attack, which I will relate in detail.

On March 4th his temperature was normal; he passed forty ounces of urine, which was quite normal in every respect.

March 5th was a clear but very cold day, and he went into the yard between nine and ten a.m. for twenty minutes. At *10.30 a.m.* he began to shiver; temperature 98°. At *10.45* he passed urine, which contained a trace of albumen, and gave a faint hæmoglobin reaction. At *11 a.m.* his temperature was 100°. At *11.15* he passed two ounces of urine, which looked like pure blood; this was at once examined microscopically, and showed no blood corpuscles, but a few round cells, much granular matter, granular casts, and oxalates. The spleen was distinctly enlarged, reaching two inches below the costal border. *11.30 a.m.* temperature 105°; five minutes later he vomited. *12 noon.* Temperature 106°; pulse 124; respiration 44. *12.30 p.m.* Temperature 105°; sweating profusely; complained of rushing noises in his head as soon as he began to get warm, and pain in his back when the attack came on, but these have now left him. *1 p.m.* Temperature 105°; copious faintly acid sweat; complains again of pain in the back. *2 p.m.* Temperature 104°; urine like tawny port, faintly acid, 1015; deposit under the microscope does not differ from that previously described; tested for indican with HCl and $CaCl_2$ gives a rose pink colour, *i.e.*, indigo red. *3 p.m.* Temperature 102°; pulse 116. *4 p.m.* Temperature 101°. *5 p.m.* Temperature 100°. *6 p.m.* Temperature 100°; has stopped sweating about half an hour: no pain in the back; pulse 88; difficulty in commencing to pass water; urine amber, 1014, faintly acid; deposits a brownish cloud, which gave the hæmoglobin reaction, and, under the microscope, was composed of granular matter. *6.30 p.m.* Temperature 99°. *9 p.m.* Temperature 98°. *11 p.m.* Temperature 98·5°; urine pale amber, 1017; deposits a mucous cloud, acid; gives hæmoglobin reaction.

March 6th. *8.10 a.m.* Temperature 98°; pulse 80; urine 1023, acid, amber; very little granular deposit; slight reaction with guaiacum; spleen cannot be felt. The urine passed later in the day contained only a faint trace of albumen; bowels confined.

March 7th. Urine free from albumen; complains of headache; tongue foul; bowels open.

He had another slight attack before leaving hospital, and was finally discharged on March 31st, but has been in hospital since.

Hayem has found free hæmoglobin in excess in the blood serum, while the well known icteroid colouring of the skin and conjunctivæ which sometimes appears, supports the view that the hæmoglobin is set free in the blood before it appears in the urine.

Hæmoglobinuria is due to rapid intravascular destruction of blood corpuscles. Hunter suggests that in pernicious anæmia the corpuscular lysis occurs in the portal system, but that in hæmoglobinuria it takes place in the peripheral parts.

Hæmoglobinuria occurs in many diseases, *e.g.*, malaria, septicæmia, and from the effects of many poisons, *e.g.*, chlorate of potash, naphthol, arseniuretted hydrogen, sulphuric acid; in all of these cases it is no doubt caused by the direct action of a poison on the blood corpuscles.

In paroxysmal hæmoglobinuria the primary cause is cold, acting on the blood in the vessels of peripheral parts, *e.g.*, hands, feet, ears and nose, where the circulation is sluggish. There is a nervous factor co-operating which probably acts by slowing the circulation in certain parts.

Blood can generally be recognised in the urine by the eye, even when present in small quantities. It is characteristic of blood that its solutions are dichroic, appearing red by reflected and green by transmitted light. But acid urine soon changes the bright red colour into a dirty brown (methæmoglobin), so that urine which has remained in the bladder mixed with blood is smoky brown or porter-coloured, according to the amount of blood present.

The *microscope* is undoubtedly the best means of detecting corpuscular blood in the urine. A drop of urine should be taken up by a pipette and placed on a glass slide, covered with a thin cover glass, and examined with a lens of at least two hundred and fifty diameters magnifying power.

If traces of blood only are present the lowest stratum

CLINICAL EXAMINATION OF THE URINE. 135

Fig. 29. Blood discs: *a*, Discs preserving their bi-concave shape; *b*, Discs swollen by imbibition of water, shedding their hæmoglobin and becoming colourless.

of urine should be examined after standing some time. The corpuscles undergo many changes, swelling up so as to lose their biconcave form, or shedding their hæmoglobin, by which they alter in shape, appear vacuolated, and ultimately colourless. (*Fig.* 29.) Such colourless discs may possibly be confounded with discoid oxalates and torulæ, but both these are smaller, while the latter contain bright nuclei and are generally oval.

The Guaiacum Test.—This depends upon the ozone-carrying power of hæmoglobin. It is generally performed by adding one or two drops of fresh tincture of *guaiacum* to about a drachm of urine, shaking the mixture, and filling in about half a drachm of ozonic ether, which is a solution of *peroxide of hydrogen* in *sulphuric ether*. If hæmoglobin be present a blue colour appears. Another method is to dip strips of blotting-paper in the tincture and dry them. These are used by touching them with a drop of urine and a drop of ozonic ether successively. *Turpentine* which has been exposed to the air always contains ozone, and may be substituted for ozonic ether, but cannot be depended upon so well. Unfortunately this test is not quite perfect. The urine of patients containing iodide of potassium gives a blue colour though no blood be present, while I have often observed that corpuscles are to be seen with the microscope when the chemical report says "no blood reaction." With respect to this want of delicacy two cautions may be given. The guaiacum tincture should not be kept longer than two months; and, secondly, the urine to be tested should be drawn up with a pipette from the lower stratum, just as it would be for microscopical observation, as the corpuscles naturally sink to the bottom of the glass.

It is stated that fibrin possesses the power of decomposing peroxide of hydrogen, giving a blue colour with the guaiacum test. I have not been able to obtain this result with well-washed fibrin, but in any case fibrin is not likely to be met with in urine apart from hæmaturia.

The *spectroscopic* examination of blood in solution is easy, and may be made by holding the vessel containing the urine between the source of light and an ordinary pocket spectroscope applied to the eye, when two dark bands will be visible between Fraunhofer's lines D and E in the yellow and green of the spectrum; if the colouring matter is converted into methæmoglobin, a band in red will appear in addition to the other two. When the blood is present in an insoluble state the urine must be filtered, and the filter paper with the deposit upon it digested in *alcohol* and *ammonia*. This fluid should be examined as in the other case, but if little blood be present the faintest possible shadow in the orange of the spectrum may alone be visible; on adding *ammonium sulphide* to the fluid the two bands will show themselves, but disappear when the fluid is shaken with air, to reappear on standing.

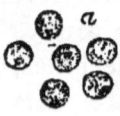

Fig. 30. *a*, Pus corpuscles; *b*, cleared by acetic acid, showing two, three, and four nuclei.

Pus may be found in the urine whenever inflammation affects any part of the urinary tract, but it is most commonly met with in catarrh of the urethra (*gonorrhœa*), of the bladder (*cystitis*), or catarrh of the pelvis of the kidney (*pyelitis*). In the second condition the urine is generally alkaline and triple phosphates are found with the pus, in the latter the urine is generally acid and the deposit commonly contains oxalates. Pus is best detected by the microscope (*Fig.* 30), the round leucocyte-like cells are easily recognised; by adding acetic acid they become clear and show two or more nuclei; sometimes they are opaque from infiltration with micro-organisms, when acetic acid will fail to clear them.

CLINICAL EXAMINATION OF THE URINE. 137

When pus is present in quantity it effervesces with *ozonic ether*, but this is not a delicate reaction. On the addition of *liq. potassæ* the liquor puris becomes ropy, even when a small quantity is present.

Pus when present in quantity in the urine forms a cream coloured deposit, which is not readily diffused into the supernatant fluid.

Casts have been fully described in a preceding chapter (*Vide* Chap. III. p. 37).

They are moulds of the tubules of the kidney and are of three kinds: (1,) *Blood* casts, formed of red corpuscles stuck together by fibrin; (2,) *Hyaline* casts, originating in various ways: (*a*,) From fibrin; (*b*,) From a proto-

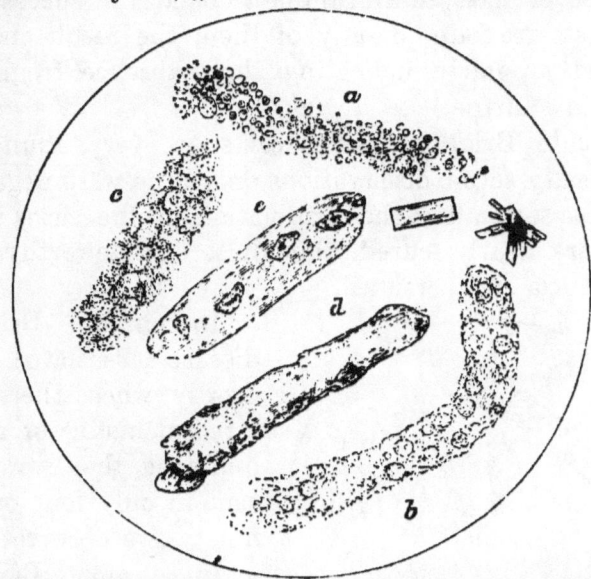

Fig. 81. *a*, Blood cast; *b*, Epithelial cast composed of small round cells; *c*, Epithelial cast formed of desquamated and fatty epithelium; *d*, Granular hyaline cast; *e*, Hyalo-epithelial cast.

plasmic exudation from the renal epithelium in the early irritative stages of acute nephritis or in chronic inflammation; (*c*,) From the colloid degeneration of desquamated epithelial cells. The second is probably the most common type; such casts are slender, while the third kind are broad and have the same meaning as epithelial casts; (3,) *Epithelial* casts, formed either by the packing together of desquamated and fatty epithelium, or by masses of round cells derived by proliferation from the epithelium of the tubules, indicate a high degree of active inflammation of the renal parenchyma (*Figs.* 31 and 32).

To find casts use a half-inch objective, with a good light. Take up a drop of the deposit or lowest stratum of the urine with a fine pipette, and place six drops on as many slides. (Very suitable pipettes are sold as "biological pipettes," made of thick glass with a bore of about one-sixteenth of an inch.) Cover each drop with a thin cover glass, and examine the slides in succession; if no casts are found on any of them the result may be regarded as fairly indicating their absence from that specimen of urine.

In acute Bright's disease casts are very abundant. Out of sixty-seven observations only nine were negative, and these were all at the termination of the cases when they were nearly cured, though it is noteworthy that albuminuria still persisted.

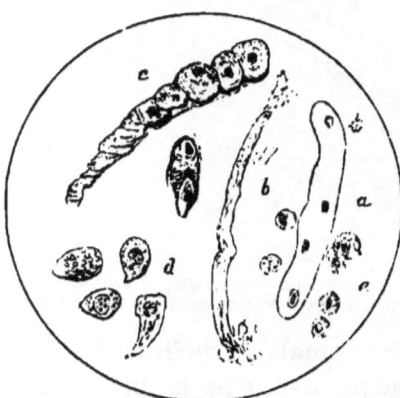

Fig. 32. *a*, Slender hyaline cast; *b*, Mucous cylinder; *c*, Hyaline and epithelial cast; *d*, Pear-shaped epithelial cells from pelvis of kidney; *e*, Epithelium from tubules.

In chronic Bright's disease associated with dropsy, where there was a recent history of acute nephritis, they were absent in only four out of ninety-five observations.

In chronic Bright's disease with little or no dropsy they were also very constant, being present in sixty-eight out of seventy-seven observations; but they were not abundant. In cases of contracting kidney seen in the outpatient room, and diagnosed as such from other signs and symptoms, casts were present in sixteen out of twenty-five observations.

Epithelium of various kinds occurs in the urine. Large squamous cells from the vagina and urethra of women, and the bladder of both sexes, are common. The epithelium from the male urethra is columnar. Pear-shaped or tailed cells may come from Cowper's glands, Littré's glands, the prostate, the ureter, and pelvis of the kidney.

Small round cells derived by proliferation from the renal epithelium may be seen in acute Bright's disease. (*Figs.* 32 and 33.)

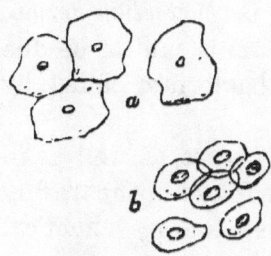

Fig. 33. *a*, Large squamous epithelium from vagina; *b*, Epithelium from bladder.

It is not always possible to say whence given cells are derived; they often have undergone fatty degeneration or infiltration by microbes, or are broken up into fragments.

Epithelial cells must be looked for with a good quarter-inch glass.

Micro-organisms are not present in normal urine, as it may be kept for months in a warm place without undergoing any change; but Kannenberg states that he has found spheroid and rod-shaped forms in the urine of healthy persons, though much more abundantly in all acute diseases. Lustgarten and Mannaberg also found organisms in the urine of healthy persons, for the urethral mucus of eight healthy persons contained ten different kinds of microbes (in four cases bacilli and in six micrococci). Of these ten varieties, two were especially notable, one being a bacillus giving the staining reactions of the *bacillus tuberculosis*, the other a micrococcus indistinguishable from the *gonococcus* or microbe of gonorrhœa. They found that it was necessary to take the precaution of drawing off the urine by a catheter in order to obtain a secretion which on cultivation was free from organisms.

They found numerous *streptococci* (organisms arranged in chains) in the carefully drawn off urine of three cases of acute Bright's disease, which disappeared on the decline of the disease. They did not succeed in obtaining a pure culture of these organisms.

Undoubtedly many organisms find their way into the urine after it has left the body; *torulæ* are not uncommon, especially in urine containing sugar. These *torulæ* are the sporules of the common moulds (*penicillium glaucum* and *aspergillus niger*), and of the yeast plant (*saccharomyces cerevisiæ*). Putrid urine swarms with bacteria, of which the best known is *bacterium termo*. The ammoniacal decomposition of urine is said to be due to the action of a minute spherical bacterium called by Cohn *micrococcus urcæ*.

In morbid conditions of the urinary tract, as well as in many specific diseases, organisms are found in the freshly passed urine which do not give rise to any chemical changes in it; one of the earliest known was *sarcina ventriculi*, which appears in the urine in certain cases of vesical catarrh. Roberts has described three cases of *bacteruria* associated with bladder troubles.

Tubercle bacilli have been found in the urine in tubercular diseases of the urinary organs, and of the epididymis (Rosenstein). Bouchard found micro-organisms in the urine of typhoid fever, puerperal fever, measles, erysipelas, dysentery, osteomyelitis, diphtheria, and phthisis.

Berlioz found the typhoid bacillus in the urine of two out of fourteen cases, in one instance on the twentieth day of the disease. In two cases of pneumonia and five of erysipelas his results were negative.

In animals inoculated with charbon he observed the passage of bacteria into the urine in two cases accompanied by hæmoglobin. The passage was found to be facilitated by cantharidin poisoning, by which the kidneys were irritated. He also observed the passage into the urine of the *micrococcus tetragonus*, the *bacillus pyocyaneus*, and the *pneumococcus* of Fraenkel. He believes that these appearances always denote a morbid localisation in the urinary apparatus.

Schweiger found that organisms injected into the renal vein or artery soon appeared in the urine, and conversely when injected into the pelvis of the kidney could be found by cultivation in the blood. He attempted to determine where the passage takes place, but was unable to detect any *in transitu*.

In many conditions micro-organisms have been observed in the kidneys *post mortem* when they have not been found in the urine, *e.g.*, *micrococcus diphtheriæ*, the micrococcus of *ulcerative endocarditis*, *microsporon septicum*, Friedländer's *pneumococcus*, etc.

The method of examining urine for micro-organisms is the following. Place a drop of the deposit, or of the urine on a cover glass which has been cleansed with nitric acid and distilled water. Place another cover glass upon the drop and press the two together, then separate them by sliding one over the other. Allow the films of fluid to dry, then holding the covers with a pair of forceps pass each two or three times quickly through the flame of a spirit lamp or Bunsen's burner. Stain them with methyl violet solution.

Saturated alcoholic solution of methyl violet	11 parts.
Aniline water	110 ,,
Absolute alcohol	10 ,,

They may be left in this solution several hours. Wash with alcohol for three minutes, then in a solution of ten parts iodine, twenty parts iodide of potassium, and three thousand parts of distilled water, until the dark blue violet is replaced by a dark purple red. Wash in alcohol till most of the colour is removed. The covers may be mounted at once on glass slides with a little Canada balsam.

This is Gram's method as given by Woodhead and Hare, to whose book readers are referred who want information about special stains and methods of cultivation.

The accompanying woodcut shows various forms of micro-organisms (*Fig.* 34).

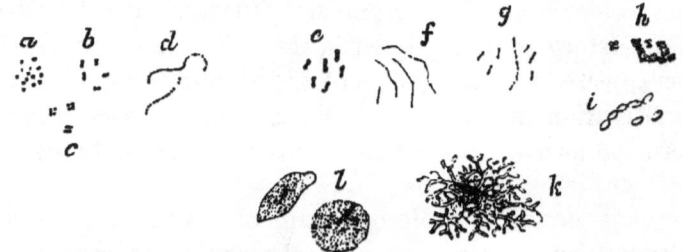

Fig. 34. *a*, Micrococci; *b*, Diplococci; *c*, Cocci in fours; *d*, Streptococci, or cocci in chains; *e*, Bacterium termo; *f*, Bacillus subtilis; *g*, Bacillus tuberculosis; *h*, Sarcinæ ventriculi; *i*, Torulæ; *k*, Mycelium; *l*, Zoogloea.

Ferments are stated to be present in normal urine. *Trypsin,* the proteolytic ferment of the pancreas which occurs in normal urine, is said by Mya and Belfanti to be replaced by pepsin in Bright's disease.

Holvotschiner and Breusing have found that there is an amylolytic ferment present which is capable of converting starch into erythro- and achröo-dextrines, but not farther according to the second of these two observers.

Béchamp's "nephrozymose," an albuminous body in urine with the power of converting starch into sugar, is said by Leube to be a compound of albumen with some amylolytic ferment.

The relation of the proteolytic ferments to peptonuria has been already alluded to.

There is no reason to think that the amylolytic ferment, even if capable of converting starch into sugar, is ever responsible for the occurrence of paradoxical glycosuria, as there is no carbo-hydrate in the urine upon which it could exercise its powers.

BIBLIOGRAPHY.

ANDERSON (E. C.). Leucin and Tyrosin in fresh urine. "Brit. Med. Jour.," 1887, II., p. 673.

AUFRECHT. Die diffuse Nephritis und die Entzundung in allgemeine. "Cent. f. d. Med. Wiss.," 1878, p. 337.

BAGINSKY (A.). Ueber Acetonurie bei Kindern. "Arch. f. Kind.," Bd. IX., 1887-8, p. 1.

BERLIOZ. Clinical and Experimental Researches on the passage of Bacteria into the Urine. Paris: O. Doin, 1887.

BREUSING (R.). Ueber das " Starke Ümwandelnde " Ferment im menschlichen Harn. "Virchow's Archiv," Bd. CVII., p. 186.

BROUARDEL (P.). L'Urée et la Foie. Paris, 1877.

BULL (G. C.). A Preliminary Report on the Presence of Acetone in the Urine. "Birm. Med. Review," XVII., 1885, p. 211.

CAMERER. Quoted by Salkowski and Leube, *op. cit.*, p. 80.

COOK (E. A.). Production and Excretion of Uric Acid. "Brit. Med. Jour.," 1888, I., p. 245.

——— Influence of Drugs on Urea and Uric Acid *Ibid.*, pp. 857, 1060.

FLUGGE. Beiträge zur Hygiene. Leipzig, 1879, p. 117.

FRANCOTTI. Non-parasitic Chyluria. "Ann. de la Soc. Méd.- Chir. de Liége," Sept., 1883.

GARROD (A. B.). Lumleian Lectures on Uric Acid. "Brit. Med. Jour.," 1888, I., p. 495.

HAIG (A.). The Relation of a certain Form of Headache to the Excretion of Uric Acid. "Med. Chir. Trans.," LXX., p. 355.

HARRIS (T.). The Relative Value of the more commonly employed Tests for Albumen in the Urine. "Med. Chron.," II., 1885, p. 459.

HAYCRAFT (J. B.). A New Method for the Quantitative Estimation of Uric Acid. "Brit. Med. Jour.," 1885, II., p. 1100.

HERMAN (G. E.). A Case of Eclampsia of Pregnancy, with Observations on the State of the Renal Function. "Brit. Med. Jour.," 1887, II., p. 1334.

——— ——— Bright's Disease during Pregnancy. *Ibid.*

HUNTER (W.). An Investigation into the Pathology of Pernicious Anæmia. "Lancet," 1888, II., pp. 555, 608.

JAFFE. Zur Lehre von den Eigenschaften und der Abstammung der Harnpigmente. "Virchow's Archiv," Bd. XLVII., p. 405.

JOHNSON (G.). On Picric Acid as a Test for Albumen and Sugar in the Urine. "Brit. Med. Jour.," 1883, I., pp. 504, 515.

KIENER and KELSCH. Les Altérations Paludéennes du Rein. "Arch. de Phys.," 1882, I., p. 278.

KIRK (R.). Report on a New Acid found in human Urine which

darkens with Alkalies (Alcaptonuria). "Brit. Med. Jour.," 1888, II., p. 232.

LANDOIS (L.). A Text-book of Human Physiology. Stirling's Translation. 1885.

LANGHANS (T.). Ueber die Veränderungen der Glomeruli bei der Nephritis nebst einigen Bemerkungen über die Entstehung der Fibrincylinder. "Virchow's Archiv.," Bd. LXXVI., p. 85.

LECORCHÉ. Traité du Diabète. Paris, 1877.

LEGG (J. WICKHAM). On Paroxysmal Hæmaturia. "St. Barth. Hosp. Rep.," X., 1874, p. 71.

LE NOBEL (C.). Ueber einige neue chemische Eigenschaften des Acetons und verwandter Substanzen und deren Benutzung zur Lösung der Acetonuriefrage. "Arch. f. exp. Path.," Bd. XVIII., p. 6.

LEPINE and AUBERT. Contribution à l'étude de la Secrétion urinaire. "Compt. Rend. de la Soc. de Biol.," Jan. 9, 1888.

LUSTGARTEN and MANNABERG. Ueber die Mikroorganismen der normalen, männlichen Urethra und des normalen Harnes, mit Bemerkungen über Mikroorganismen im Harne bei Morbus Brightii acutus. "Viertel. f. Derm. und Syph.," 1887, p. 905.

MACMUNN (C. A.). Researches into the Colouring Matter of Human Urine, with an Account of the Separation of Urobilin. "Proc. Roy. Soc.," No. 206, 1880.

MAGUIRE (R.). The Darkening in Colour of certain Urines on exposure to the Air. "Brit. Med. Jour.," 1884, II., p. 808.

MONVENOUX. Les Matières grasses dans l'Urine. Paris.

NAUWERCK (C.). Beiträge zur Kenntniss des Morbus Brightii. Ueber Morbus Brightii bei croupöser Pneumonie.

NEUBAUER and VOGEL. A guide to the qualitative and quantitative Analysis of the Urine. 4th edition. London, 1863.

NOEL-PATON (D.) The commoner methods for the estimation of Urea in Urine. "The Practitioner," XXXVI., 1886, p. 168.

———— Report on the relationship of the formation of Urea and Uric Acid to the Secretion of Bile. "Brit. Med. Jour.," 1886 I., pp. 377 and 433.

ORD (W. M.). Urinary Crystals and Calculi. "Med. Chir. Trans.," LVIII., 1875, p. 165.

POSNER (C.). Studien über pathologischen Exudatsbildungen. "Virchow's Archiv," Bd. LXXIX., p. 311.

PROUT (W.). On the nature and treatment of Stomach and Renal Diseases. 5th edition. London, 1848.

RALFE (C. H.). A practical treatise on Diseases of the Kidneys and Urinary Derangements. London, 1885

RALFE (C. H.). Phosphatic Diabetes, "Lancet," I., 1887, pp. 411, 462.

RANKE. "Virchow's Archiv," 1862; quoted by Salkowski and Leube, *op. cit.*, p. 80.

RIBBERT (H.). Ueber die Eiweissausscheidung durch die Nieren. "Cent. f. d. Med. Wiss.," 1879, p. 47, and 1881, p. 17.

ROBERTS (W.). A practical treatise on Urinary and Renal Diseases. 4th edition. London, 1885.

——————— On the chemical constitution and physiological relations of the amorphous Urate Deposit. "Med. Chron.," March, 1888, p. 441.

ROSSBACH. Ueber Chylurie. "Verhandl. d. VI. Congr. f. innere Med.," Wiesbaden, 1887, p. 212.

RUBNER. Quoted by Salkowski and Leube, *op. cit.*, p. 80.

SALKOWSKI and LEUBE. Die Lehre vom Harn. Berlin. 1882.

SCHULTZEN. "Virchow's Archiv," 1863; quoted by Salkowski and Leube, *op. cit.*, p. 80.

SCHWEIGER (F.). Ueber das Durchgehen von Bacillen durch die Nieren. "Virchow's Archiv," Bd. CX., p. 255.

SEEGEN (J.). Ueber Diabetes mellitus mit Rücksicht suf die neugewonnenen Tatsachen über Zuckerbildung im Tierkörper.

SMITH (W. G.). Note on a case of peculiar Albuminous Urine. "Brit. Med. Jour.," 1880, II., p. 847.

——————— On the occurrence of Protocatechuic Acid in Urine. "Dub. Jour. Med. Sci.," June, 1882.

——————— On the nature of the Phosphatic Precipitate obtained upon heating Urine. "*Ibid.*," July, 1883.

——————— On the conditions affecting precipitation of Phosphate of Lime from Urine by heat. "Brit. Med. Jour.," 1883, II., p. 68.

STRAUSS and GERMONT. Des lésions histologiques du Rein chez le Cobaye à la suite de la Ligature de l'Uretère. "Arch. de Phys.," Vol. IX., 1882, p. 386.

THOMSON (R. STEVENSON). Scarlatinal Albuminuria, and the pre-albuminuric stage. "Med. Chir. Trans.," Vol. LXIX., p. 97.

VOIT. Quoted by Salkowski and Leube, *op. cit.*, p. 80.

VOORHOEVE (N. A. J.). Ueber das Entstehen der sogenannten Fibrincylinder. "Virchow's Archiv," Bd. LXXX., p. 247.

WEISSGERBER and PERLS. Beitr. zur. Kenntniss der Entstehung der sog. Fibrincylinder nebst Bemerk. über Mikrokokkenanhäufungen in der Niere bei Blutstauung. "Arch. f. exp. Path.," Bd., VI., p. 113.

WINDLE (B. C. A.). On the Ferric Chloride Reaction in Urine. "Liverpool Med. Chir. Jour.," July, 1884.

Woodhead and Hare. Pathological Mycology. Edinburgh, 1885.

Yvon and Berlioz. L'Urine normale. " Revue de Méd.," Sept. 1888.

Ziegler and Nauwerck. Beiträge zur pathologischen Anatomie und Physiologie. Jena, 1887.

Section III.—BRIGHT'S DISEASE.

Chapter IX.
HISTORY—CLASSIFICATION—ETIOLOGY.

ALTHOUGH Van Helmont regarded the kidneys as the seat of the causation of dropsy, and the discovery of Cotunnius that the urine of dropsy was coagulable by heat was published as early as 1770, there can be no doubt that the whole honour of establishing the true relations of dropsy and albuminuria to disease of the kidneys belongs to the great physician and pathologist of Guy's Hospital, RICHARD BRIGHT.

Blackall, of Exeter, would perhaps have forestalled him had he enjoyed equal opportunities for making *post mortem* examinations; but, in the fourth edition of his book, published in 1825, Blackall shows himself to be entirely ignorant of the local causes of dropsy, assigning to it a constitutional origin, and inclining to the opinion that albuminuria was due to *the elimination of the dropsical fluid by the urinary passages.*

Bright's *Reports of Medical Cases*, published in 1827, present a striking contrast in the definite solidism of his pathology to this vague humoralism. He distinctly ascribes albuminuria and dropsy to the altered anatomical condition of the kidneys, and he figures accurately the changes in the kidneys just as they are recognised by us to-day.

In his papers in the first volume of *Guy's Hospital Reports*, published in 1836, Bright added a great deal of clinical and pathological information.

He had learnt that dropsy might be slight or altogether absent, and that albuminuria might require looking for. He described the various complications, the inflammations of serous membranes, hæmorrhages, apoplexies, convulsions, blindness, and coma; he drew attention to the frequency of cardiac hypertrophy, and suggested an explanation which still holds its ground.

He recognised the importance of alcohol and exposure to cold as etiological factors, while his views on prognosis were truer and more liberal than those which afterwards became current.

But on the actual nature of the pathological process his ideas were cramped by the contemporary state of pathological doctrines. Just as in these days every disease is ascribed to a *microbe*, in Bright's time everything was regarded as due to a *deposit*. Laennec called tubercle and cirrhosis of the liver deposits. So Bright thought the various anatomical types of Bright's disease were stages in the evolution of a deposited material, and the hard granular kidney was the ultimate result of the process.

Rayer, in 1839, correctly described the inflammatory nature of the lesion, and some years after (1851) Frerichs explained the different anatomical types by his doctrine of three stages: (1,) Hyperæmia with exudation; (2,) Fatty degeneration of the exudation; and (3,) Absorption of the exudate with atrophy of the organ.

Johnson, in his "Diseases of the Kidney," published in 1852, made two important additions to the subject: (1,) He showed that the chief seat of acute inflammation was the tubules, and that in the sequel the tubular epithelium underwent necrosis, desquamation, and fatty degeneration; (2,) He differentiated the small red kidney as a type which occurred independently of acute inflammation.

In 1858 Virchow gave an academical form to these

discoveries by describing three conditions: (1,) Parenchymatous inflammation, originating in the tubular epithelium; (2,) Interstitial inflammation, originating in the connective tissue; and (3,) Amyloid degeneration, originating in the blood vessels.

Amyloid, waxy or lardaceous degeneration had been previously recognised by Rokitansky and Johnson, but only as a complication, not as a distinct type.

Rosenstein, in 1860, found Virchow's academic classification too rigid, and described: (1,) Hyperæmia; (2,) Catarrhal nephritis; (3,) Diffuse nephritis; and (4,) Amyloid degeneration.

Grainger Stewart, in 1868, recognised three forms of Bright's disease: (1,) Inflammatory, having three stages: (a,) Of inflammation; (b,) Of fatty transformation; (c,) Of atrophy. (2,) Waxy or amyloid, also having three stages: (a,) Affecting vessels only; (b,) Transudation into tubules; (c,) Atrophy. (3,) Cirrhotic or contracting.

In 1872 Gull and Sutton introduced an entirely new doctrine to explain the pathogenesis of the contracting kidney. They re-discovered the thickening of the vessels which had been described by Johnson, and announced the existence of a generalised affection of the small arteries and capillaries, to which they gave the name of *arterio-capillary fibrosis,* and of which they maintained the kidney lesion was merely a pronounced local expression; the vascular degeneration having led to atrophy of the surrounding tissues.

But the doctrine of interstitial nephritis received a new impetus from the careful work of Kelsch, published in 1874, and in succeeding years it was endorsed by Bartels, Charcot, and Grainger Stewart.

It had never been accepted by Johnson, Moxon or Roberts in this country, or by Rosenstein, who declared "That there could not properly be said to be either strictly parenchymatous or strictly interstitial nephritis," both tissues being affected, and "that large white and small red kidneys were alike the result of *diffuse* inflammation."

Weigert gave great support to this reaction by a paper published in 1879, in which he maintained that the parenchyma and the stroma are affected in all cases of chronic Bright's disease, and that pure parenchymatous nephritis exists only as an acute disease. These views met with the concurrence of Bamberger, and were supported by the experiments of Grawitz and Israel, who found that artificially induced nephritis was followed indifferently by the small red or large white kidney.

In 1880 I expressed the results of my own histological observations in the following words: " The small red and large white kidney, and all the intermediate varieties, are the result of inflammation which affects all the tissues, but varies very greatly in intensity. The parenchyma being the most highly organised tissue, suffers most in proportion to the intensity of the inflammation. The large pale kidney is the result of prolonged or repeated severe inflammation; on the other hand, the small red kidney indicates an inflammatory process of prolonged duration but of minimum intensity; and the intermediate varieties correspond to all the different degrees of intensity possible between the two extremes. The fact of the existence of *an indefinite number of intermediate or mixed forms* between the two typical varieties of the large white and small red kidney is a strong argument in favour of the doctrine of unity."

Since that time there has been no conspicuous attempt to revive the doctrines of Virchow, and the opinion that the lesion is a diffuse nephritis in all forms of chronic Bright's disease has steadily gained ground.

In his recent valuable book on kidney diseases Ralfe fully accepts this view.

One of the latest writers who have investigated this

subject is Snyers, who formulates his conclusion in the following terms: "After a microscopical examination made without any preconceived opinions, the absolute separation and the formulated opposition of parenchymatous and interstitial nephritis cannot be sustained."

CLASSIFICATION.

In the preceding sketch of the history of the doctrine of Bright's disease, I have recorded the failure of successive attempts to classify the various clinical forms upon an anatomical basis.

I propose to adopt an etiological classification, which I hope will be found to adapt itself to the facts of clinical observation as well as to pathology. The following are the proposed divisions:

(1,) Febrile Nephritis;
(2,) Toxæmic Nephritis;
(3,) Obstructive Nephritis.

The first division includes all those cases of acute or chronic nephritis occurring as a result of acute or chronic febrile diseases. The nephritis is directly dependent upon the fever process, hence its name. Pathologically it includes most cases of acute parenchymatous nephritis and of chronic fatty kidney.

The second division includes the great group of chronic Bright's disease due to lithæmia, and is specially associated with the small red granular kidney. It probably depends upon irritation of the kidneys by the excessive elimination of poisons of which uric acid is the type. It includes the acute nephritis of acute gout, of poisoning by animal, vegetable or mineral poisons, and certain cases of primary acute nephritis usually attributed to chill, but in which there is probably an already existing dyscrasia. Such cases are met with occasionally in individuals who get drunk on beer and lie out all night.

The third division includes all cases dependent upon

obstruction to the outflow of urine. They occur commonly in males as a consequence of stricture or enlarged prostate, in females from pressure on the ureters caused by pregnancy or pelvic diseases.

In each of these classes we may meet with the urinary and other symptoms of acute or chronic nephritis, while the *post mortem* appearances may be those of acute nephritis, of chronic fatty kidney, or of contracting kidney.

Lardaceous or *waxy* degeneration is not made a special group because it is only when associated with chronic nephritis that it deserves to be called Bright's disease. General lardaceous degeneration affecting liver, spleen, kidneys, intestinal mucous membrane, lymphatic glands, heart, etc., has no other title to be called Bright's disease than the occurrence of albuminuria, and even that is not always constant (FÜRBRINGER). Whatever may have been the case twenty years ago, it cannot be maintained now, and certainly will not be admitted here, that albuminuria and Bright's disease are synonymous.

The lardaceous kidney of most authors is chronic nephritis as it occurs in chronic pyrexial diseases, *e.g.*, long standing suppurations, phthisis &c., in which lardaceous degeneration may occur just as it may in any form of chronic Bright's disease.

Its relation to suppuration is by no means so constant as has been taught; for out of sixteen cases of chronic nephritis occurring in these circumstances, collected from the pathological registers of the General Hospital, lardaceous degeneration was present in two only.

I admit that lardaceous disease is less common in Birmingham than in London or Edinburgh; this is a striking and curious fact; but the disease is by no means unknown, and its relative rarity makes its true relation to Bright's disease more apparent.

This is after all only reverting to the older opinion of Rokitansky and Johnson.

GENERAL ETIOLOGY.

PREDISPOSING CAUSES.—There can be no doubt in the minds of pathologists of the remarkable frequency of Bright's disease in this country, although the Registrar-General's returns and even our hospital registers fail to give an adequate account of it.

Out of a total number of deaths in England and Wales for the year 1884, amounting to 530,828, only 6,297, or 1·1 per cent., are returned as having been registered under the various terms which are included by Bright's disease. In this district, the West Midland, there were only six hundred and ninety out of a total of 56,938 deaths, or 1·2 per cent.

In London alone during that year there were 11,000 deaths in persons over fifty years of age, of which only seven hundred and sixty were registered as Bright's disease; but Mahomed has told us that in the *post mortem* registers of Guy's Hospital, he found out of a hundred and fifty cases over fifty years of age, sixty-two instances of chronic Bright's disease, a proportion of one in 2·42, so that instead of seven hundred and sixty, there should have been 4,546 cases of Bright's disease.

On examining the pathological registers of the General Hospital for ten years, from 1875 to 1884, out of 1,213 deaths at all ages there were no less than three hundred and eighty-three cases of acute and chronic Bright's disease, giving a percentage of 31·5 or a proportion of about one in three.

But the Registrar-General's returns shew that there has been of late years an increase in the number of cases registered as due to this cause, the increase being in the less obvious form of chronic Bright's disease; for while in 1875, out of a population of twenty-two millions, 3,841 cases were registered under Bright's disease, of which number nine hundred and seventy-eight were described as '*nephritis*'; in 1884, with a population of twenty-seven millions 6,297 cases were registered as Bright's disease, of which number only 1,177 were described as '*nephritis*.' The increase of population was twenty per cent., of cases of *nephritis* eighteen per cent. (less than the population), and of Bright's disease sixty-four per cent!

This result is in all probability due to the fact that the profession is beginning to look for latent cases of chronic Bright's disease and has learnt to recognise them better than was formerly the case.

I think we may assume the power of recognition to be a uniform factor in the comparison of one registration district with another, for the purpose of ascertaining whether there is any great difference in the frequency of the disease in particular portions of this kingdom. The following are the results of the figures worked out as percentages for the year 1884:—

All England	1·1 per cent.
London	1·4 ,,
South Eastern	1·4 ,,
South Midland	1·4 ,,
Eastern	0·9 ,,
South Western	1·4 ,,
West Midland	1·2 ,,
North Midland	1·0 ,,
North Western	1·0 ,,
York	1·0 ,,
Northern	0·7 ,,
Wales	1·4 ,,

This district, the West Midland, is very little above the average. The influence of great towns does not seem to be very decided, looking at the high figure of Wales, and the low figures of all the northern districts.

I am not inclined to regard this table as giving information of a very trustworthy character. It is very remarkable that the eastern district, in which stone is so common, stands so low. It is contrary to what we know of the etiology of Bright's disease, and is probably explained by the fact that the form of the disease set up by uric acid is the type most likely to be overlooked.

The prevalence of Bright's disease is said to be equally great in Holland, Denmark, Scandinavia, and on the shores of the Baltic, which share with us a common predisposing cause, a COLD MOIST CLIMATE.

In such an atmospheric medium as we live in the functions of the skin are habitually depressed, and an undue share of the work of elimination is thrown upon the kidneys. A general pathological law links together excessive function and proneness to disease. In tropical climates we have a converse illustration in the great prevalence of skin diseases.

But, in all probability, it is not simply an excess of work that the kidneys are called upon to do. Garrod states that suppression of perspiration is followed by increased acidity of the urine, and from this we may infer diminished alkalinity of the blood, leading to the accumulation of uric acid in the system.

Semmola believes that the chief effect of chill is an alteration of the blood albuminoids resulting in renal irritation, hyperæmia and albuminuria, which is followed by general nutritive changes and anatomical alterations in the kidneys. In effect he thinks that the altered blood albuminoids act like egg albumen when injected into the veins, becoming readily diffusible through the glomerular walls and appearing in the urine. Thus he makes albuminuria the first link in the chain of sequences, to be followed by structural alteration in the kidney.

Such a theory needs proof, and of this none is offered. On the contrary, the analogy of egg-albuminuria is a weak one, for it leaves no ill effects on the kidneys (SNYERS). Moreover, Tizzoni found that albumen from the urine of Bright's disease when injected into the veins of animals did not cause albuminuria, and there are many cases on record or known to me of albuminuria that has persisted for years without giving rise to any other symptom of Bright's disease.

I may quote a case of persistent albuminuria which occurred in a young gentleman who was subject to attacks of paroxysmal hæmoglobinuria. The albuminuria first appeared after an attack early in 1878. Late in that year he was sent to Italy for the winter as a case of Bright's disease. I saw him after his return when the albumen still existed, and I know it continued until 1882. He had another attack of hæmoglobinuria in 1883, but I have not seen him since 1882 professionally, although I hear of him very often, and know him to be in excellent health at the present time (1888).*

Sex.—It is commonly stated that Bright's disease is more common in males than in females. The pathological register of the General Hospital shows its proportion to all deaths in either sex to be in males 43·4, in females 40·6. Bamberger finds no difference.

In women there are special causes, such as pregnancy, uterine and ovarian tumours, and pelvic inflammation, which are important and common factors; but against these we may place the greater exposure of men, their habits in food and drink, and the frequency of cystitis, stricture, etc.

Age.—Acute Bright's disease is much more common in children than adults on account of its relation to acute specific diseases, especially scarlatina. Chronic Bright's disease is more frequent after middle life.

The mortality from Bright's disease shows a progressive increase as age advances; a fact permitting the just inference that chronic Bright's disease is much more fatal than the acute attack.

---

* See also cases quoted Chapter I. (p. 11).

GENERAL ETIOLOGY.

	Under 5	5 to 15	15 to 25	25 to 35	35 to 45	45 to 55	55 to 65	65 to 75
Males	168	145	189	290	469	644	724	606
Females	140	123	174	284	412	482	543	488

Heredity.—There is an undoubted tendency in Bright's disease to attack members of the same family and to appear in successive generations.

In a case of contracting kidney in a lad, seen in consultation with Dr. J. A. Lycett, of Wolverhampton, his father and two paternal uncles had died of Bright's disease. Striking instances have been published (MEIGS, KIDD.) This may be attributed to its relation to gout and lithæmia. There is probably an inherited vice of structure or function; a thin ill-developed skin is undoubtedly a transmissible peculiarity which has come under my observation in this connection. The habit of constipation, upon whatever it depends, is also a feature common to families, as are also the habits of eating and drinking to excess. The following is an example of three cases of Bright's disease in one family.

CASE 8.*—William E., aged seventeen, waggoner, admitted Dec. 11th, 1887, with dropsy of face, legs and abdomen. Four weeks before he had got very wet; this was followed by shivering, lumbar pain, swelling of face, legs and belly, with diminished urine. In a week the pain went away, the urine increased in amount, and the dropsy diminished.

He could remember no previous illness, and his statement was confirmed by his mother; he had never had scarlatina. His work exposed him to wet and cold.

His father had died of kidney disease, aged thirty-two;

---

* Recorded by Mr. Teichelmann, clinical assistant.

and an uncle was at present under treatment in hospital for the same condition.

Patient was a slightly-built lad, with a little œdematous swelling of face and legs. T. 99·2°; P. 72; R. 19; Tongue clean; bowels regular; no ascites; liver and spleen normal.

Heart's apex in fifth interspace internal to vertical nipple line; a systolic murmur at apex. Pulse 72, regular, full. Urine 48 oz.; sp. gr. 1010; acid; contains albumen and blood.

Dec. 14th. Better; no œdema of legs; urine 60 oz.

Dec. 15th. Urine 50 oz.; sp. gr. 1018; acid; pale straw colour; mucous cloud; urea 242 grains *pro die;* albumen a cloud; a trace of blood; hyaline, epithelial and granular casts, red and white blood corpuscles and renal epithelium.

Dec. 16th. No œdema of face now.

Dec. 26th. Urine 52 oz.; sp. gr. 1014; acid; opaque, pale straw colour; mucous deposit; urea 251·68; a cloud of albumen; a few granular casts; red and white blood corpuscles and a few renal epithelia.

Jan. 4th. Allowed to get up.

Jan. 5th. Urine 46 oz.; sp. gr. 1020; acid, dark straw colour; mucous deposit; urea 344 grains *pro die;* a cloud of albumen; a very little blood; no casts or renal epithelium.

Jan. 10th. Abundant hæmaturia, causing headache; sent back to bed. No œdema.

Jan. 14th. Urine 52 oz.; sp. gr. 1015; neutral; pale straw colour; urea 274 grains *pro die;* a faint cloud of albumen; no blood; a few hyaline casts.

He was discharged on Jan. 25th, without any further relapse.

Social State.—Bright's disease attacks all classes; but there are occupations that are specially liable to it; these are, the manufacture and sale of alcoholic drinks, brewers, distillers, publicans, and the like; workers in lead, file casters, glass cutters, workpeople in white lead factories, painters, lapidaries, &c.; those specially exposed to cold or damp—well sinkers, miners, &c., and to extreme changes of temperature—furnace men, iron workers, stokers, &c.

Diet.—Our habits as to food and drink constitute an etiological factor of great importance.

We are great eaters of butcher's meat, not only the great source of urea and uric acid, but which also contains a large quantity of acid salts, by which the alkalinity of the blood is reduced with results already detailed.

We aggravate this evil by drinking beer, containing a large amount of acid, chiefly in the form of acetic acid.

Finally, we consume an enormous quantity of alcohol in other forms, which is almost wholly eliminated by the kidneys, and which certainly leads to structural disease of the liver.

Previous Diseases.—Chronic heart, lung, and liver diseases, by leading to impairment of assimilation, and especially to the imperfect fulfilment of the great urea-forming function, are undoubtedly remote causes of Bright's disease.

So, too, habitual constipation predisposes by favouring a dyscrasia caused by imperfect intestinal elimination, as well as perhaps by the absorption of animal alkaloids, formed in the putrefactive processes which take place when food remains too long in the alimentary canal.

EXCITING CAUSES.—Of the *exciting causes* of Bright's disease, there are three great groups, whose efficiency is established by an overwhelming mass of evidence. The best known of these clinically is the great group of *acute* and *chronic febrile diseases*.

Acute diseases, such as scarlatina, pneumonia, typhoid

fever, variola, diphtheria, measles, varicella, tonsillitis, and acute rheumatism (E. WAGNER), give rise to acute nephritis; while chronic diseases, such as phthisis, chronic septicæmia, and malarial fever cause chronic nephritis.

Gaucher attributes the nephritis to the irritation caused by the increase of extractives (kreatin, kreatinin, leucin, tyrosin, xanthin, and hypoxanthin) in the blood which are eliminated by the kidneys, and which he has proved by experiments on animals can set up nephritis. This suggestion links this with the next great group, that of *poisons*.

The most interesting and important of these are those substances which are formed in the body, and of which we may take *uric acid* as the type (MURCHISON). They are normal products of disassimilation, but under certain circumstances are produced in excess. These circumstances have been already alluded to. They are circumstances of climate, of individual conformation, of occupation, and above all, of habits as to food and drink; also the presence of pre-existing diseases, especially of the heart and liver. Poisoning of this kind is usually a very chronic and insidious process, often giving rise to no symptoms until the destruction of the kidneys has advanced so far that the dyscrasia, which hitherto has been only the *cause* of the renal disease, now becomes intensified by the failure of the kidneys to eliminate the impurities from the blood, and sets up all the constitutional disturbances known as uræmia. In other cases the disease is discovered earlier, before the kidneys are so far destroyed, when the destructive process may be checked by treatment designed to limit this auto-intoxication.

Many other poisons are known to be efficient causes of nephritis, but are not of great clinical importance. The following is a list of them:—

Animal Poisons :—Cantharides

ETIOLOGY.

Vegetable Poisons :—Oxalic acid
　　　　　　　　　Opium (Nauwerck)
Mineral Poisons :—Arsenic
　　　　　　　　　Mercury
　　　　　　　　　Iron
　　　　　　　　　Manganese
　　　　　　　　　Cobalt
　　　　　　　　　Nickel
　　　　　　　　　Zinc
　　　　　　　　　Lead (?)
　　　　　　　　　Sulphuric Acid
　　　　　　　　　Nitric Acid
　　　　　　　　　Hydrochloric Acid
　　　　　　　　　Carbolic Acid
　　　　　　　　　Ammonia
　　　　　　　　　Iodoform

The third group comprises all *obstructive causes* from pressure on or diseases in the urinary passages. In males the principal conditions are stricture, enlarged prostate, and tumours and diseases of the bladder, *e.g.*, cystitis, calculus; in females pregnancy, uterine and ovarian tumours, pelvic inflammations, and bladder diseases.

Simple obstruction sets up a chronic inflammatory process tending to contraction of the kidney. Kidneys which are undergoing this process of obstructive atrophy are very liable to attacks *of acute interstitial inflammation* originating in the medulla and spreading to the cortex. Such attacks may be set up by *cystitis*, or by even a slight traumatism such as may be caused by the *passage of a catheter* for the first time.

BIBLIOGRAPHY.

BAMBERGER (H. von). Ueber Morbus Brightii und seine Beziehungen zu anderer Krankheiten. 1879.

BARTELS (C.). Ziemssen's Cyclopædia of the Practice of Medicine. Eng. Ed., 1876, Vol. XV.

BLACKALL (J.). Observations on the nature and cure of Dropsies. London, 1824.

BRIGHT (R.). Reports of Medical Cases. London, 1827.

—————. Cases and observations illustrative of Renal Disease accompanied with the secretion of Albuminous Urine.

—————. Tabular view of the Morbid Appearances occurring in one hundred cases in connection with Albuminous Urine. "Guy's Hosp. Reports," Vol. I., 1836, pp. 338 and 380.

CHARCOT (J. M.). Leçons sur les Maladies du Foie et des Reins. Paris, 1877.

COTUNNIUS, —., quoted by Darwin. "Zoonomia," Vol. I., 1794, p. 310.

FRERICHS (F. T.). Die Bright'sche Nierenkrankheiten und deren Behandlung. Braunschweig, 1851.

FÜRBRINGER (P.). Zur Diagnose der amyloiden Entartung der Nieren. "Virchow's Archiv.," Bd. LXXI., Heft 3.

GAUCHER (E.). Recherches expérimentales sur la pathogénie des Néphrites par Auto-intoxication. "Revue de Méd.," 1888, No. 11, p. 885.

GULL and SUTTON. On the pathology of the morbid state commonly called Chronic Bright's disease with Contracted Kidney. "Med. Chir. Trans.," Vol. LV., 1872, p. 273.

JOHNSON (G.). On Diseases of the Kidneys. London, 1852.

—————. Lectures on Bright's disease. London, 1873.

KELSCH (A.). Revue critique et recherches anatomo-pathologiques sur la Maladie de Bright. "Arch. de Phys.," Vol. VI., 1874, p. 722.

KIDD (J.). The Inheritance of Bright's disease of the Kidney. "Practitioner," Vol. XXIX., 1882, p. 104.

MAHOMED (F. A.). Some of the clinical aspects of Chronic Bright's disease. "Guy's Hosp. Reports," 3rd S., Vol. XXIV., p. 363.

—————. Chronic Bright's disease and its essential symptoms. "Lancet," 1879, I., p. 46.

—————. Chronic Bright's disease without Albuminuria. A thesis for the degree of Bachelor of Medicine of the University of Cambridge, read June 10th, 1881.

MEIGS (A. J.). Clinical observations on Albuminuria, based upon a study of sixty-two cases seen in private practice. "Boston Med. and Surg. Jour.," Vol. CVII., 1882, p. 409.

MURCHISON (C.). Lectures on Diseases of the Liver. 2nd Ed. London, 1877, p. 573.

NAUWERCK (C.). Akute Nephritis bei Opiumsergiftung, Beiträge zur Kenntniss des Morbus Brightii.

RALFE (C. H.). A practical Treatise on Diseases of the Kidneys and Urinary derangements. London, 1885.

RAYER (P.). Traité des Maladies des Reins et des altérations de la Secrétion Urinaire. Paris, 1839—44.

ROSENSTEIN (S.). Pathologie und Therapie der Nierenkrankheiten. 2nd Ed., Berlin, 1870. 3rd Ed., 1886.

SAUNDBY (R.). The Histology of Granular Kidney. "Path. Soc. Trans.," 1880, p. 148.

SEMMOLA. Nuove contribuzioni alla patologia ed alla cura del Morbo di Bright. "Med. Contemp. Napoli." 1886, III., pp. 449 to 467.

SNYERS (P.). Pathologie des Néphrites chroniques. Bruxelles, 1886.

STEWART (GRAINGER). Bright's diseases of the Kidneys. Edin. 1868. 2nd Ed., 1871.

—————————. On certain morbid states of the Kidney. Address in the Section of Medicine, Brit. Med. Assoc., 1878.

TIZZONI (G.). Alcuni esperimenti intorno alla patogensi dell' Albuminuria. "Gaz. degli Ospit. Milano," Vol. VI., 1885, p. 12.

TRAUBE. Ueber den Zusammenhang von Herz- und Nierenkrankheiten. Berlin, 1856.

VAN HELMONT. Oriatrike or Physick refined. London, 1362, p. 507.

VIRCHOW (R.). Cellular Pathology. London, 1858.

WAGNER (E.). Beiträge zur Kenntniss des chronischen Morbus Brightii. "Deutsch. Arch. für Klin. Med.," Bd. XXVII., p. 218.

WEIGERT. Die Bright'sche Nierenkrankung vom pathologisch-anatomischen Standpunkte. "Volkmann's Sammlung Klin. Vorträge," Serie VI., pp. 162-3.

Chapter X.

GENERAL ANATOMY OF THE KIDNEY.

The shape of the kidney, in health, is a flattened ovoid, larger and rounded above, flattened and pointed below. Its colour is dark-red with a smooth surface. The average length is 4·4 inches, the breadth two inches, and the thickness one inch.

The usual weight is about five and a half ounces in the male, and five ounces in the female. At the end of this chapter a table of weights at different ages is appended. The outer border is convex and rounded, while the inner is concave and traversed by a longitudinal fissure,—the *hilum*, in which are situated the renal vein, ureter and renal artery. The hilum leads into a hollow in the kidney called the *sinus*, in which the vessels lie before penetrating the renal substance.

The kidney is enveloped by a thin tough fibrous capsule which is continuous with the outer fibrous coat of the ureter. This capsule is attached to the renal substance by fine processes of connective tissue and by blood vessels.

Under the capsule there is a layer of un-striped muscular fibres.

On section the kidney is seen to be made up of two portions: (1,) an outer light brown portion, the *cortical layer*, of friable consistence and fairly homogeneous appearance; and, (2,) an inner darker portion composed of a series of cones, the pyramids of Malpighi, whose apices are turned towards the sinus, the *medullary layer*. This inner layer is divided by anatomists into two parts, —a *boundary zone* consisting of the broader portions of the pyramids, and a *papillary portion*, constituted by the projecting apices.

The relative proportions of these three divisions are according to Klein

Cortical Layer	3·5
Boundary Zone	2·5
Papillary Portion	4·0

The structure of the cortex is slightly granular from the presence of the Malpighian bodies, while it is marked by lighter lines at right angles to its surface.

The *medullary rays*, or pyramids of Ferrein, are conical bundles of straight tubules running from the cortex into the medulla.

Between the medullary rays are the interlobular vessels, the Malpighian bodies, and the convoluted tubules, together forming the *labyrinth*.

The medulla is distinctly striated by alternating light and dark lines arranged fan-wise, the handles of the fans being at the apices of the pyramids. The whitish lines are bundles of straight tubes, the darker lines blood vessels.

In the fœtus the kidney is distinctly lobulated, each lobule consisting of a pyramid of Malpighi with its corresponding cortical portion. In the adult this lobulated condition persists only rarely, but the connection between each pyramid and the uriniferous tubules and blood vessels of its cortical portion always remains (*Fig.* 37).

The *uriniferous tubules* take origin in Bowman's capsules, membranous expansions of their extremities, measuring $\frac{1}{250}$ of an inch in diameter. They pursue a tortuous course, effect junctions with one another, and ultimately form large collecting tubes which open on the apices of the papillæ. Each tubule is formed of a homogeneous basement membrane lined by epithelial cells, with an axial space or lumen for the passage of urine.

Fig. 35 represents the course of a uriniferous

tubule. Connected by a narrow neck (2) with Bowman's capsule (1) is the *proximal convoluted tubule* (3) which ultimately leaves the labyrinth and passes down the medullary ray into a pyramid. Before leaving the

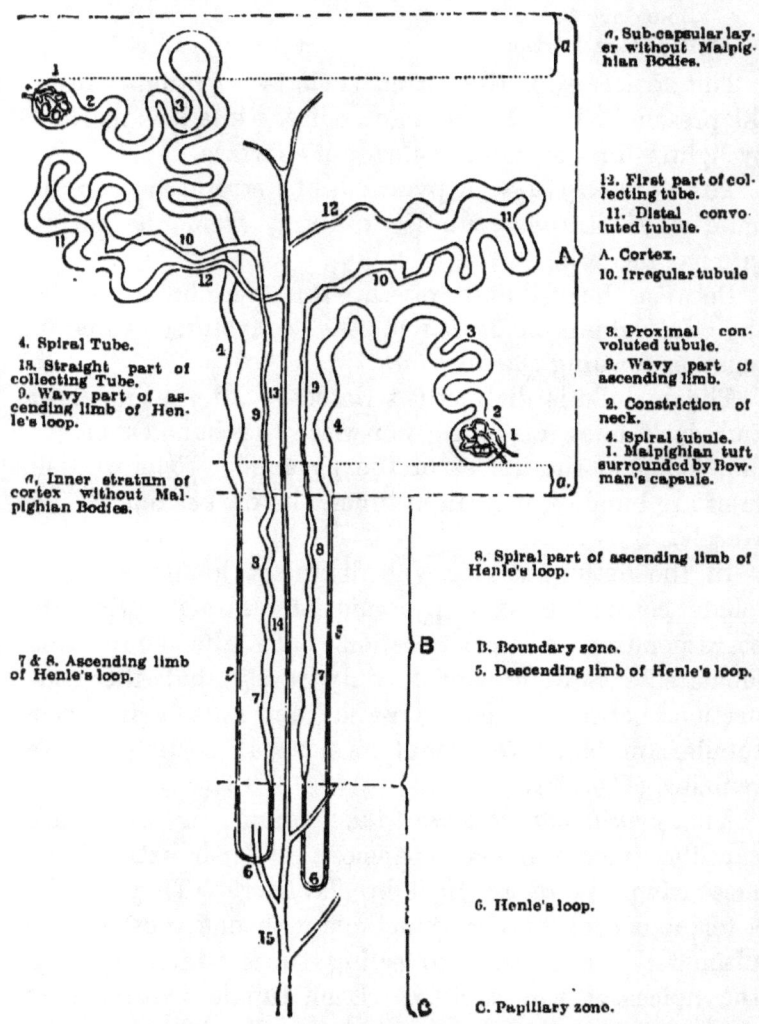

Fig. 35. Diagram of course of two urinary tubules (after Landois and Stirling.)

labyrinth its convoluted appearance becomes less pronounced, and it is called a *spiral tubule* (4); descending, it narrows very much, finally turning at Henle's loop (6), the descending and ascending portions being called the *descending* and *ascending limbs of Henle's looped tube*. The ascending limb is at first spiral, then wavy, and on reaching the cortex it re-enters the labyrinth as an *irregular tubule* (10),—a short irregular portion which soon becomes wide and convoluted forming the *distal convoluted tubule* (11), which after pursuing a tortuous course again enters a medullary ray as a *collecting tube* (12); this collecting tube passes down into a pyramid, is joined by others and opens on the surface of the apex of a papilla.

The convoluted tubules are lined by a large granular epithelium arranged in a single layer. The cells are composed of two parts, an inner containing a spherical nucleus, and an outer part which has a fibrillated appearance from the presence of rods placed at right angles to the axis of the tube.

The cells appear continuous, no separation between those being ordinarily visible, adjacent cells interlocking by tooth-like projections on their surfaces.

Similar epithelium is found in the *spiral tubule*, and the *ascending limb* of Henle's loop.

In the *descending limb* the epithelium is clear and flattened with a bulging nucleus.

In the *irregular tubule* the cells somewhat resemble those of the convoluted tubes, but they are shorter, the nuclei are oval and the rods are coarser and better defined.

The *collecting tubes* are lined by large clear cubical or columnar epithelium with distinct nuclei.

Malpighian bodies are small spherical structures composed of Bowman's capsules, already alluded to, which enclose capillary tufts, and are pierced by *afferent* and *efferent* vessels.

They are situated upon the extremities of horizontal

branches of the interlobular arteries and are about $\frac{1}{250}$ of an inch in diameter. Within the capsule each afferent vessel breaks up to form a capillary tuft, an arrangement of vascular loops supported by connective tissue and covered by a layer of endothelium (*Fig.* 36, GL).

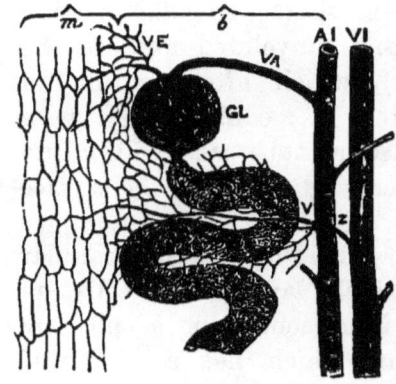

Fig. 36. Diagram of arrangement of blood-vessels in the renal cortex (after Tyson). *m*, region of the medullary ray; *b*, region of the convoluted tubules; VI, interlobular vein; AI, Interlobular artery; VA, vas afferens; VE, vas efferens; GL, Bowman's capsule containing capillary tuft; VZ, venous radicle of interlobular vein.

The inner surface of the capsule is likewise covered by a lining of flattened cells which are continuous with and probably modifications of the tubular epithelium.

Blood vessels. — The renal artery as it enters the hilum divides into four or five branches which are invested by sheaths derived from the fibrous capsule. They pass inwards between the bases of the papillæ to form arches with one another in the boundary zone.

From these arches the *interlobular arteries* arise; they ascend towards the surface of the cortex between the medullary rays, furnishing lateral horizontal branches,—the afferent twigs to the Malpighian bodies (*Fig.* 36, AI). Other branches are supplied to the matrix of the kidney and to the capsule, the latter anastomising with branches of the supra-renal, phrenic, and lumbar arteries.

On leaving the Malpighian body the efferent vessel, which in structure resembles an artery, breaks up into a capillary meshwork surrounding the tubules; from this meshwork venous trunks arise which terminate in the interlobular veins (*Fig.* 36, VI). The interlobular

GENERAL ANATOMY OF THE KIDNEY. 169

veins originate just below the capsule as the *venæ stellatæ*. They anastomose freely in the boundary zone, and accompany the arteries to the sinus, where they finally unite to form the *renal vein*.

From the same arches in the boundary zone, formed by the branches of the renal artery after their entrance into

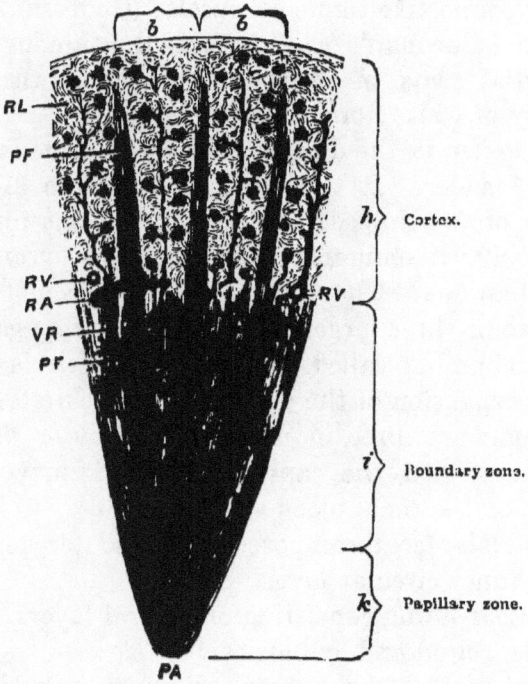

Fig. 37. Diagram of a pyramid of Malpighi with corresponding cortex (after Landois and Stirling.) PF, Pyramids of Ferrein or medullary rays; RA, Branch of renal artery; RV, Branch of renal vein joined by interlobular vein; VR, Vasa recta; PA, Papilla; b, b embrace the bases of renal lobules.

the kidney, other vessels arise called the *vasa recta* (*Fig.* 37, VR), which enter the pyramids and pass between the tubules towards the apices.

Some vasa recta are said to arise from efferent vessels which lie near the medulla or even from the union of the capillaries of the medullary rays (HUSCKE).

In their course through the pyramids they form capillary meshworks around the tubules. From these capillaries venous radicles arise, the *venæ rectæ*, which open into the venous trunks in this region.

The *connective tissue* of the kidney consists mainly of an intertubular stroma, which in the papillæ is broad and fibrous, but in the cortex is more delicate, and is said to be composed almost entirely of branching and anastomosing connective tissue corpuscles (GOODSIR).

Around Bowman's capsules and continuous with the adventitial coats of the blood vessels is also a small quantity of white fibrous tissue.

The *ureter* is the duct by which the urine is conveyed to the bladder. It takes origin from the kidney in a number of cup-shaped tubes, called *calices* or infundibula, each of which embraces the apex of a pyramid, or of more than one. The several calices become united to form two or three larger tubes which unite again in the dilated chamber called the pelvis, which is a funnel-shaped expansion of the upper end of the ureter.

Its coats are three in number. An outer *fibrous* coat continuous with the capsule of the kidney and the sheaths of the renal blood vessels.

A *muscular* coat composed of an outer longitudinal and an inner circular layer.

A *mucous* lining consisting of several layers of epithelial cells, round and cylindrical.

Lymphatics of the kidney originate in wide meshed plexuses in the capsule and in large spaces which are more numerous between the convoluted tubules of the cortex, but are also present between the straight tubules.

Large lymphatics provided with valves pass out of the kidney at the hilum while others emerge through the capsule.

The *nerves* are derived fom the renal ganglia and lesser splanchnic nerve. They are gangliated, and contain both medullated and non-medullated fibres. Their mode of termination is unknown.

GENERAL ANATOMY OF THE KIDNEY.

Table of weights of the kidneys at different ages (taken from Boyd's tables).

Age.	Sex.	Average weight both kidneys.	Age.	Sex.	Average weight both kidneys.
		oz.			oz.
1 to 2	Male	2·55	30 to 40	Male	11·35
	Female	2·4		Female	10·34
2 to 4	M.	3·33	40 to 50	M.	10·89
	F.	3·14		F.	8·8
4 to 7	M.	4·05	50 to 60	M.	9·1
	F.	4·26		F.	8·55
7 to 14	M.	6·58	60 to 70	M.	8·83
	F.	5·75		F.	8·28
14 to 20	M.	9·34	70 to 80	M.	10·68
	F.	9·09		F.	7·68
20 to 30	M.	11·57	Upwards of 80	M.	8.25
	F.	10·17		F.	6·86

Chapter XI.
FEBRILE NEPHRITIS.

This division includes all cases of acute nephritis where the inflammation is set up in consequence of the fever process in the course of some acute disease. It also includes all those cases of chronic nephritis occurring as the direct result of chronic fever processes in phthisis, prolonged suppurations, etc., and also those occurring as the sequel to acute nephritis of febrile origin.

Etiology.—The following acute diseases are known to cause acute nephritis: Scarlatina, diphtheria, pneumonia, enteric fever, exanthematic typhus, remittent fever, malarial fever, variola, measles, varicella, whooping cough, mumps, acute rheumatism, tonsillitis, erythema nodosum, erysipelas, septicæmia, ulcerative endocarditis, cancrum oris, and anthrax. In the greater number of cases the disease is not recognised at the time.

Phthisis, tubercular diseases of bones and joints, malaria, syphilis (?), and chronic septicæmia, set up chronic nephritis.

The disease is directly due to the fever process, and depends either upon the high temperature or upon some toxic substances formed in the blood in consequence of the fever, and eliminated by the kidneys.

Specific micro-organisms have been found in the kidneys in cases of diphtheria, pneumonia, septicæmia, and ulcerative endocarditis, but it is reasonable to conclude that they are not essential to the production of the nephritis; they are found in the kidney because they exist in the blood, and it is the natural channel for their elimination.

That nephritis has been caused experimentally by the injection of micro-organisms (micrococcus pyocyaneus,

CHARRIN and BOUCHARD) only proves that this organism may set up an acute febrile process in which the kidney suffers.

The same explanation may be given of the cases of so-called *mycotic* nephritis described by Litten, Bamberger, Aufrecht, and Mircoli.

Non-specific micro-organisms have been frequently found in the nephritis of acute diseases, *e.g.*, variola, scarlatina, and cancrum oris, but they have no pretension to any special etiological significance.

ACUTE FEBRILE NEPHRITIS.

MORBID ANATOMY.—In the majority of cases during the first week the kidneys undergo no change in size shape, or colour.

After a week, if the disease persists, the kidneys become more or less enlarged, pale in colour, or mottled red and white. On section, the swelling and pallor or mottling are seen to be in the cortex, the outspread pyramids retaining their red colour.

Histology.—The following is a brief summary of the changes in acute nephritis (under six weeks) with a magnifying power of 450 diameters:—

Blood vessels.—The blood vessels are nearly always full of blood clots, especially below the capsule and in the medulla. They are sometimes greatly dilated and filled with clot. Occasionally thrombi formed of leucocytes are seen in the vessels of the cortex.

Glomeruli.—The capillaries are often full of clots; the endothelial nuclei on the tuft are always distinct and often increased in number. Sometimes the capsular space contains blood clot, at others there is some local accumulation of cells from inflammatory exudation.

Tubules.— The epithelium of the convoluted tubes and Henle's loops is generally vacuolating (*Fig.* 38), that is, undergoing a process of atrophy by shedding its protoplasm, leaving only the basal part with the nuclei. The cavity of the tubules becomes filled with spherical droplets

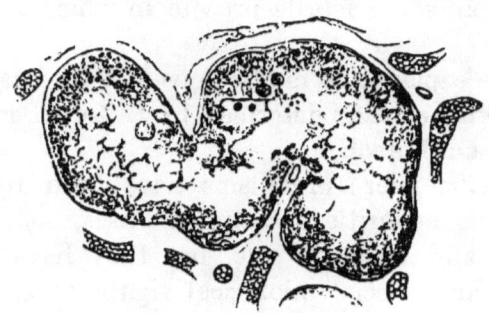

Fig. 38. Convoluted tubule in acute nephritis. The epithelium is cloudy, swollen and ruptured.

and granular matter derived from the epithelium (*Fig.* 39). In some cases the epithelium appears to be undergoing simple necrosis, breaking down into granular material. The epithelium of the collecting tubes appears to undergo no pathological change, but the cavities of many tubes

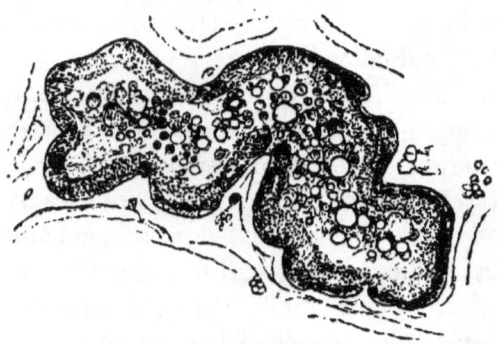

Fig. 39. Convoluted tubule in acute nephritis. Epithelium granular, partially destroyed, only the basal portion remaining. The lumen of the tubule is filled with protoplasmic droplets and granular matter.

are filled with droplets and granular matter forming casts.

Stroma.—In scarlatinal nephritis the stroma of the cortex and medulla is swollen and often filled with round cells. In other forms there are only a few round cells, and in the majority of cases it is only slightly swollen and hyaline.

This description shows the affection to be mainly an acute catarrhal inflammation affecting the epithelium of the convoluted tubes. In scarlatinal kidneys, interstitial and glomerular affections are often pronounced.

SYMPTOMS AND COURSE.—As a rule patients with acute nephritis are pale; they often complain of headache; pain referred to the loins is less common than might be supposed. There is usually complaint of weakness; the tongue is generally clean, and the bowels almost invariably confined. The last circumstance is important in connection with the influence which constipation was proved by Mahomed to have on the development of scarlatinal nephritis. It is probable that there are many cases of acute nephritis which remain latent and get well, and many more would do so but for such an accident as a chill to the surface, or constipation.

Dropsy.—More or less œdema is commonly present in acute nephritis, but it is less constant than has been stated. It is notably absent in the majority of cases occurring in the course of acute diseases. The great exception to this rule is scarlatina, in which the presence of dropsy draws attention to a nephritis which in many other acute diseases passes unrecognised.

The amount of dropsy varies very much, from mere puffiness of the eyelids to general anasarca with effusion into the serous sacs.

The eyelids are the most constant seat, next the lower extremities, then the trunk, scrotum, etc.

Œdema of the conjunctiva is a common symptom of Bright's disease, giving the eye a peculiar lustre.

Œdema of the lungs and glottis are complications which will be described hereafter.

Temperature.—The onset of acute nephritis does not seem to be attended by any marked inflammatory fever.

The temperature in cases seen early in their course is sometimes 100°, but after a few days it becomes normal or subnormal, and does not rise except on account of some complication.

Heart.—The heart is liable to attacks of acute softening or myocarditis, in which it may become rapidly dilated, and this dilatation may be followed by rapid hypertrophy. As a rule the left ventricle is alone affected. The enlargement of the heart leads to displacement of the apex beat outwards, and a systolic murmur may be audible over it. In other cases the first sound at the apex is reduplicated, and the second sound in the aortic area accentuated. These alterations are the result of the increased pressure in the aortic system, causing delay in the systole of the left ventricle followed by rapid and abrupt closure of the aortic semilunar valves.

Pulse.—The rate of the pulse is at first quick, ninety to a hundred, but falls after the first few days to seventy or eighty, not seldom to sixty. The pulse is usually moderate in volume and not perceptibly hard, but the sphygmograph shows a characteristic curve of high tension (*Fig.* 40).

Fig. 40. High tension pulse tracing of acute nephritis. The curve above the line *a b* is the full tidal wave.

In explanation of this discrepancy I would repeat my former remark that this curve is not really a measure of blood pressure, but of the duration of the tidal wave, which depends upon two factors, the peripheral obstruction and the force of the ventricle; whenever the former is too much for the latter the curve of high tension is produced, though in reality the ventricle may be weak and the intra-vascular pressure below normal.

Ophthalmoscopic appearances.—The fundus of the eye presents no change in acute Bright's disease of febrile origin; except perhaps œdematous swelling in uræmic amaurosis (*vide* p. 81).

Urine.—The *quantity* of urine is always reduced, at any rate for the first few days; it is generally under 30 oz., often under a pint in twenty-four hours.

The *density* is generally over 1020, often higher; in fact some of the densest urines are met with in acute scarlatinal nephritis, a sp. gr. of 1065 having been recorded in one instance (W. G. SMITH).

The *reaction* is almost invariably acid, unless affected by drugs. The *colour* and *translucency* depend mainly upon the amount of blood present; thus the urine may vary from being merely smoky through various shades of reddish brown to the colour of porter. If blood is present in very small quantity the urine is yellow and clear.

The *deposit* is always considerable, consisting of mucus, epithelium, casts and blood; its colour and consistence vary from white and flocculent up to chocolate and dense, the principal factor in the change being the amount of blood present.

Uric acid crystals, urates of soda, potash and ammonia, or triple phosphates sometimes occur in the deposit.

The *urea* is generally under two per cent., often little more than one per cent.; so that as the quantity of urine is small the total eliminated is usually under two hundred grains *per diem*.

Albumen is often not very abundant. A dense cloud is a fair description of the amount; more accurately from ·05 to ·1 per cent., but in severe cases it may reach one per cent. or more.

Willey has shown that acute nephritis is often found *post mortem* in scarlatinal cases which have presented no albuminuria during life.

Blood is most constantly present, generally in large, sometimes only in small amount; it often persists for long and recurs in a troublesome manner. The *casts* most commonly seen are blood and hyaline, the most significant are epithelial.

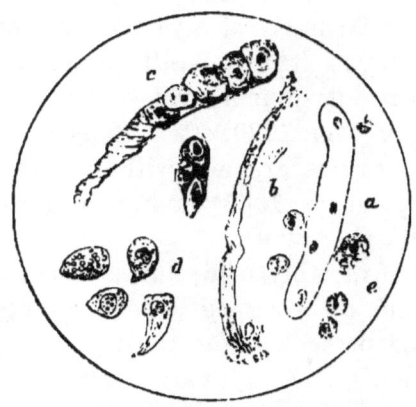

Fig. 41. *a*, Deposit in acute nephritis showing slender hyaline cast; *b*, mucous cylinder; *c*, hyaline and epithelial cast; *d*, pear-shaped epithelial cells from pelvis of kidney; *e*, epithelium from tubules.

Epithelial cells are usually abundant. The following may be taken as a typical report of the urine of an ordinary case of acute nephritis:—

Urine 10 oz.; sp. gr., 1020; acid; smoky; brownish deposit; urea, 1·2 per cent.; albumen, ·2 per cent; under microscope deposit contains epithelial, hyaline, granular and blood casts, red and white blood corpuscles and abundant renal epithelium.

DIAGNOSIS. — Even when dropsy is absent, if the urine is scanty, albuminous, and contains blood and epithelial casts, acute nephritis may be safely diagnosed. The difficulty is to distinguish between acute nephritis in previously healthy kidneys and an intercurrent attack in the course of chronic nephritis. This can be only determined with certainty by the history, but the age of the patient is suggestive. Thus in children primary disease is very common, while after forty years of age it is very rare. The presence of albuminuric retinitis or hæmorrhages indicates latent contracting kidneys, as does also, though with less certainty, the discovery of distinct enlargement of the heart early in the case. In intercurrent acute attacks the quantity of urine rapidly rises to normal or more than the normal amount.

COMPLICATIONS. — Primary acute attacks are not attended by many complications, but a mild degree of *uræmia* is often present giving rise to *headache* and *vomiting*. *Convulsions* sometimes occur even in mild cases in children, and these may be followed by *coma*. *Œdema of the glottis* is a localisation of the dropsy attended by special risk on account of its seat at the opening of the air passages.

DURATION.—It is not easy to fix the exact duration of an acute attack. The more urgent symptoms usually cease in two or three weeks under suitable treatment, but the kidneys are so far damaged that albuminuria continues as a rule for many months, during which the patient is in imminent danger of a relapse. If the symptoms persist beyond six weeks, the case is usually regarded as becoming sub-acute or chronic. Our language in this respect is a little defective, and it is plain that no very hard and fast line can be drawn.

PROGNOSIS.—The chances of recovery from an acute primary attack are decidedly favourable. Death rarely occurs during the acute stage. It is not easy to speak so confidently as to the ultimate result. A large proportion no doubt get quite well; in a certain number the kidneys remain damaged sufficiently to render them very liable to fresh attacks; while others become chronic. In all cases the utmost care is needed, and an opinion as to the ultimate result must be very guarded until months have elapsed and all trace of albumen has disappeared from the urine.

If this account of the prognosis of acute nephritis should appear to some to be too favourable, I would remind them that we recognise the disease much more commonly and in milder phases than was formerly the case. Thus Thomson's statistics show, that out of a hundred and eighty cases of scarlatina, one hundred and twelve, or 63·2 per cent., were affected with nephritis; that is to say the urine contained albumen, blood and tube casts, but only twenty-four showed any anasarca.

Of the hundred and twelve only eleven died, of whom four are stated to have died from malignant scarlatina. One died from uræmia. Of the eleven fatal cases only five showed any anasarca, though two cases in which this was noted as abundant were fatal. The case that died from uræmia had only slight dropsy.

These figures shew that the majority of cases, even of post-scarlatinal nephritis, when properly cared for, do well.

The guides to an opinion as to the probable termination of the case should be (1,) The evidence of the functional power of the kidneys as shown by the daily amount and character of the urine; (2,) The degree of dropsy present.

ILLUSTRATIVE CASES.

CASE 9.*—*Acute Bright's disease following sore throat (tonsillitis?); persistent hæmaturia.*

George T., aged twelve, schoolboy, admitted March 4th, 1887, with weakness and hæmaturia.

Eight weeks ago he had a sore throat, and after he had got over that he began to swell. His grandmother, with whom he lived, thought he had got a chill going to school. His family history was imperfect, but negative so far as it could be ascertained; his parents were not supposed to be dead. His previous health was good, but he had had scarlatina and measles when a baby. He was a fairly healthy-looking boy, with rather a puffy face; skin dry and brown; slight œdema of the legs. T. 100°; P. 78; R. 18; tongue dry, white; appetite good; no pain after food, or vomiting; bowels confined (he said they were regular); liver and spleen normal; no ascites. Heart's apex in fifth interspace inside nipple line; a systolic murmur at apex, with accentuation of the second sounds at the base. Pulse irregular; lungs normal.

---

* Recorded by Mr. Ross Jordan, Clinical Assistant.

No lumbar pain ; says he gets up in the night to make water; urine 14 oz. ; sp. gr. 1014; acid; contains albumen and blood.

March 12th. Œdema has disappeared.

March 13th. Urine 16 oz.; sp. gr. 1017, acid ; dark smoky red colour ; dark reddish brown deposit ; urea ·95 per cent. ; albumen quarter column ; epithelial granular and blood casts, red and white blood corpuscles, squamous epithelium, uric acid, and granular matter in deposit.

This condition of urine continued for two months, except that the quantity was, on an average, normal. He was not discharged till June 13th. The last urine report is as follows : Urine 35 oz. ; sp. gr. 1016 ; acid ; straw colour, slightly turbid ; reddish white flocculent deposit ; urea 2 per cent. ; a cloud of albumen ; a trace of blood ; granular and epithelial casts, red and white blood corpuscles, squamous epithelium and uric acid. He was made an out-patient, but did not attend.

CASE 10.* — *Acute Bright's disease; uræmia; recovery.*

John B., aged six, schoolboy, admitted April 3rd, 1886, with swelling of legs and face. He had been ill a week. Four years and a half previously he had scarlatina, not followed by dropsy ; two of his brothers had it at the same time.

He was a well nourished child with a pale swollen face, and considerable anasarca of the trunk and lower extremities; his eyelids were puffy. T. 98; P. 102; R. 24. Tongue slightly furred ; bowels confined ; no ascites. Liver and spleen normal. Heart's apex in fifth intercostal space just external to the vertical nipple line ; a faint systolic murmur was audible in the mitral area,

---

* Recorded by Dr Stacey Wilson, House Physician.

but the heart sounds elsewhere were loud and accentuated. Pulse small, regular, rather hard. Breath sounds rather feeble posteriorly; lungs otherwise normal.

Ophthalmoscopic appearances normal. Urine, 16 oz. in 24 hours; acid; sp. gr. 1018; in colour and appearance like beef tea; 1½ in. of brown deposit; urea, 1·3 per cent.; albumen 0·17 per cent.; blood in large amount. Epithelial, blood, hyaline, and granular casts, with blood corpuscles and renal epithelium in deposit.

He was put on milk diet, and treated by diaphoretics and diuretics.

April 12th. The œdema of the legs has disappeared.

April 17th. Hæmaturia still abundant; ordered ext. ergotæ liq. ♏v, *quartis horis.*

April 23rd. Blood reaction with guaiacum and ozonic ether feeble.

April 27th. He had an attack of uræmic convulsions which lasted an hour and a half, but not attended by loss of consciousness.

After this attack his pulse fell to 60, and was labouring, irregular and intermittent.

April 28th. Ordered solution of nitro-glycerine (1 per cent.), ♏j, *secundis horis.*

May 13th. He was much better; the pulse still intermits at long intervals; he was allowed to get up.

June 4th. He had been going on very well, when the hæmaturia recurred.

June 12th. By this time the blood was nearly gone.

July 8th. He was discharged free from œdema, and in fair health; urine 30 oz.; amber; acid; slight flocculent deposit; urea 1·7 per cent.; albumen ·05 per cent.; an occasional hyaline cast and a few blood corpuscles in the deposit.

Urine of John B., acute nephritis:—

Date.	Quantity of urine in oz.	Quantity of urea in grains.	Quantity of albumen in grains.	
April 5th ...	16	91·52	119·6	
,, 12th...	16	147·84	70·4	
,, 19th...	14	117·04	30·8	
,, 23rd...	16	112·64	14·0	
,, 28th...	20	123·2	44·0	
May 3rd...	16	105·6	7·0	
,, 12th...	24	137·28	10·5	
,, 19th...	36	237·6	15·8	
,, 25th...	30	145·2	13·2	
June 2nd...	30	198·0	6·6	
,, 7th...	30			
,, 9th...	30	118·8		Too little to estimate.
,, 16th...	22	92·4		
,, 23rd...	40	191·2		
,, 27th...	26	193·48		
July 8th...	30		7.0	Urea not estimated.

CASE 11.*—*Ulcerative endocarditis; sub-acute febrile nephritis; uræmia. Death. Autopsy.*

E. S——, female, eighteen, domestic servant, admitted July 2nd, 1886, complaining of pain at the heart and swelling of the legs and face. For the last four months she had been ailing, but her legs began to swell fourteen days before admission. She had had rheumatic fever three times in the last five years. Her previous health was good.

Father died of phthisis, aged forty-eight. Mother died of heart disease, aged forty-nine. One brother died of heart disease, aged thirteen. Two died in infancy; four are living and in good health. Patient has marked œdema of eyelids and face, slight œdema of legs and ankles; T. 100; P. 96; R. 30; tongue clean; digestion good; bowels regular; liver and spleen not enlarged.

Heart's apex beat in fourth left interspace, internal to nipple line, but dulness extends half an inch beyond the nipple in the fifth interspace. A loud systolic murmur

---

* Recorded by Dr. Stacey Wilson, House Physician.

in the mitral area; pulmonary second sound loudly accentuated. Pulse small, but not easily compressed.

Lungs normal, except some dulness with deficient breath and voice sounds at left base.

She has not menstruated for six weeks.

Has no pain or difficulty in making water.

Urine, 26 oz.; 1014, acid; straw-colour, blood-tinged; reddish deposit; urea, 1·4 per cent.; a cloud of albumen; numerous hyaline and granular, with a few colloid and epithelial casts, red and white blood corpuscles, squamous and renal epithelium, and uric acid crystals.

After admission her temperature assumed a hectic type, rising at night sometimes as high as 103° Fahr. Her urine remained much the same.

On August 10th she became drowsy, and this was followed by uræmic convulsions. Two days later the convulsions returned, coma supervened, and she died.

Autopsy performed by Dr. Crooke.

Subject emaciated; features pale and puffy; lips livid; feet and ankles swollen. Abdomen distended.

Heart, 11½ oz., pericardium universally adherent by old fibrous adhesions; both ventricles dilated, the left especially. Left ventricle contains dark semi-fluid blood clots; watery fluid blood, and colourless soft clot in right auricle and ventricle; valves in right side normal.

Aorta, rather narrow, valves normal. Extensive recent endocarditis round mitral valve extending into left auricle on its outer wall, where there is a large patch of vegetations. The mitral curtains are beset with soft greyish friable vegetations. Wall of left ventricle vascular, cloudy, mottled yellow in places, soft and friable.

Lungs congested in their lower lobes; upper lobes œdematous.

Liver 56 oz., soft, pale, with hyperæmic patches.

Spleen 13½ oz., adherent, contains a large infarct

FEBRILE NEPHRITIS.

which is breaking down, and another smaller one softening in the centre.

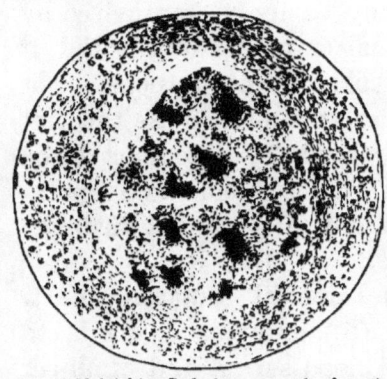

Fig. 42. Malpighian Body from case of sub-acute nephritis in ulcerative endocarditis, showing infiltration of the tuft with micro-organisms (streptococci.)

Several hæmorrhages in spleen substance, which is very vascular.

Kidneys 9 oz., capsules strip easily, leaving a characteristic greyish brown and red speckled surface. On section the cortex is increased, presenting the same finely mottled, grey brown and red coloration, mixed with a sprinkling of yellow.

The cut surface is dotted with whitish points, which are the glomeruli altered by inflammatory change. At the inferior end of the left kidney is a small wedge-shaped infarct, yellowish white, dry looking and firm, fringed with a zone of hyperæmia. The medullary cones are uniformly light chocolate brown in colour. Under the microscope the Malpighian bodies are infiltrated with micro-organisms, chiefly *streptococci* (*Fig.* 42).

TREATMENT.—The main lines of treatment are rest and warmth, aided by suitable diet, purgatives and diaphoresis.

The patient should remain in bed, in a flannel night-dress, wrapped in a blanket, or the sheets may be taken off the bed. The diet should be *milk* and farinaceous food. No beef tea (according to Masterman beef tea is analogous in chemical composition to urine except that it contains less urea and uric acid), meat extracts or jellies should be given. The bowels should be kept open by an occasional purgative (bitartrate of potash ʒss, honey or treacle ʒj), every second or third day if required. The best diaphoretic is the *hot air bath*, which should not be prolonged beyond twenty minutes. It will be found that a patient who does not sweat at all in the bath on the first few occasions will do so perfectly well afterwards. The hot air bath may easily be improvised by a spirit lamp set under a stool or cradle in the bed, if one of the cheap tin lamps sold for the purpose is not available. I very much prefer this method to the hot pack, which has lately been shown by Goodhart to be dangerous by causing pyrexia. I have had the temperature taken many times before and after the hot air bath, and have not noticed any rise.

I usually give a pint of Imperial (potassii bitart. ʒss., sacch. alb. q.s., aquæ Oj,) daily as a drink, to keep up the alkalinity of the blood serum and diminish the acidity of the urine, nor have I seen any reason to believe it harmful, in spite of the theoretical importance of potassium salts in the pathogenesis of uræmia.

Great responsibility attaches to the medical attendant with respect to the care and conduct of his patient after convalescence is reached. Undoubtedly great, sometimes insuperable difficulties are placed in his way, but it is his plain duty to point out as impressively as he can the dangers that lie ahead and the precautions needed to avoid them.

The patient must be told of the great liability there is for the disease to relapse from exposure to cold, from injudicious diet, and from the occurrence of any acute disease, however slight.

If possible he should spend the winter in a milder climate than our own, or in the mildest parts of this island. He should not be allowed to eat butcher's meat or cheese, or to drink alcohol in any form until all albumen has disappeared from his urine, and even then he should be warned to use these articles of diet very sparingly.

His dress should consist of underclothing of wool from

head to foot, summer and winter, and he should be prevented from taking part in work or exercise likely to overheat his body or risk the possibility of a chill.

CHRONIC FEBRILE NEPHRITIS.

ETIOLOGY.—As a result of neglect of proper precautions after an acute attack, often in ignorance of its existence, the patient gets a chill, and a fresh attack is lighted up which may pass into chronic nephritis. Moreover such kidneys are specially liable to be irritated by the various poisons which set up toxæmic nephritis in healthy kidneys.

In chronic febrile diseases there is a correspondingly chronic nephritis which often remains latent for a considerable part of its course.

MORBID ANATOMY.—As a rule the kidneys become larger and paler; the capsule still strips off readily, leaving a smooth shining surface. On section, the cortex is broad, white, and soft, while the pyramids are pale red and streaked with white lines. The Malpighian bodies often show the reaction of lardaceous or waxy degeneration with liq. iodi. In its most pronounced type this constitutes one form of the "large white kidney," but this anatomical variety may originate in other ways.

In some cases the kidney is no larger than normal or may be even smaller, and instances have occurred in which one kidney is large and the other small, as in the following description of a pair of kidneys copied from the *post mortem* register of the General Hospital.

Kidneys of H.—The right is enormously swollen, and weighs 14½ oz., while the left is much smaller, weighing only 6 oz., and is much lobulated. Both are pale, soft and flabby in consistence, capsules easily separable, leaving a dull white ground colour mottled with red and yellow (branny kidney). On section, the *right* or larger kidney presents large white areas in which the fatty changes (yellow branny speckling) are marked; the cortex is much increased in breadth, of a uniform grey white colour, swollen, generally opaque, but in some places semi-translucent in appearance, shewing also a fine yellow speckling of the surface, and streaked and dotted with lines and points of hyperæmia. The *left* or smaller kidney presents very similar appearances except that it is rather more vascular; no difference can be felt in its resistance to the knife, and it is as soft and doughy as the right kidney. At the bottom of the depressions or lobulations the capsule appears slightly thickened and adherent, and a small greyish semi-translucent strand appears to enter the cortex.

In other cases, especially in post-scarlatinal kidneys, in which interstitial nephritis was noted as occurring during the primary acute attack, the organs are reduced in size, with thickened and adherent capsules, and pale roughened surfaces. On section, the kidney substance is abnormally tough, while the cortex is narrowed.

This is the type which was formerly described as a third stage, in which the large white kidney had undergone atrophy, but it is now thought more probable that these kidneys have never been much enlarged, and that interstitial inflammation has been a marked feature in the inflammatory process from the beginning. It is not clear under what circumstances, apart from the clinical fact of its frequency in scarlatina, this particular form is developed. As already mentioned, Grawitz and Israel found the two types of kidney occurring in animals indifferently as the result of artificial nephritis induced by temporarily occluding the renal artery.

Histology. — The *Malpighian bodies* present hyaline change of their capillaries. The capsules frequently show marked peri-capsulitis, being swollen and broadened out by concentric bands of nucleated fibres; while their endothelium is proliferated, sometimes forming a mass of granulation tissue (*Fig.* 43,) encroaching on the tuft. The *blood vessels* are dilated and full of clot; they are often thickened by hypertrophy involving all their coats. The *convoluted tubes* are dilated, their epithelium is fatty, only the basal portion being left (*Fig.* 44). The cavities of the tubes contain fat granules and epithelial débris. The *straight tubes* of the medulla preserve their epithelium, but their cavities often contain casts made up of fatty granules and epithelial débris from the convoluted tubes. The *stroma* is everywhere swollen, nucleated and hyaline.

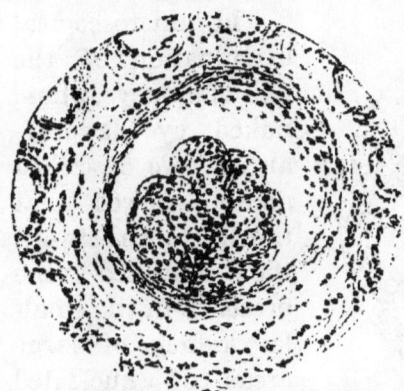

Fig. 43. Malpighian body shewing marked peri- and endo-capsulitis, with increase of the nuclei of the tuft.

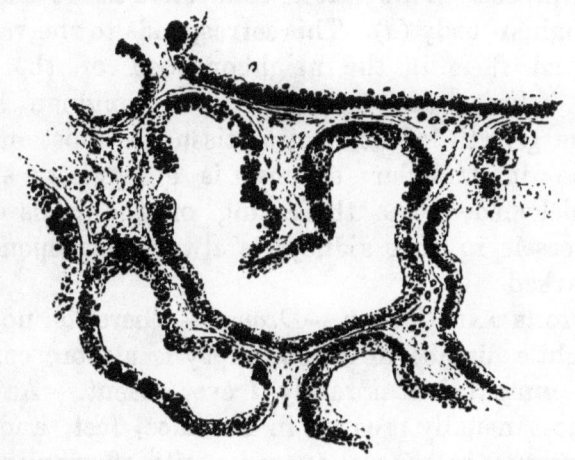

Fig. 44. Osmic acid preparation shewing marked atrophy and fatty degeneration of the epithelium of the convoluted tubules.

These changes indicate a diffuse inflammation of the glomeruli, coats of the blood vessels, and stroma, with fatty degeneration of the previously inflamed epithelium of the convoluted tubes.

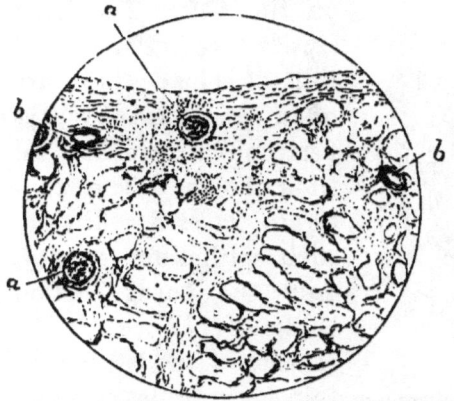

Fig. 45, Section of kidney from chronic febrile nephritis brushed out so as to shew the thickened stroma.

The microscopical examination of the pair of kidneys, whose naked eye appearances were described above, shewed that the difference between them consisted in the presence in the smaller kidney of large areas of nucleated fibrous tissue, enclosing atrophied tubules.

Fig. 45 is a preparation from such a kidney brushed out so as to remove all but the framework of connective tissue. In the centre of the section there is a wedge-shaped process of nucleated connective tissue enclosing a Malpighian body (a). This corresponds to the vascular zone, and it is in the neighbourhood of the inter-lobular arteries and around the Malpighian bodies, that the growth of connective tissue is most marked. But the intertubular stroma is everywhere swollen and thickened. The thickening of the walls of the blood vessels in such kidneys is always correspondingly well marked.

SYMPTOMS AND COURSE.—*Dropsy.*—There is no form of Bright's disease in which dropsy is a more constant accompaniment. It is rarely if ever absent. As a rule it is most usually present in the face, feet, and legs, but there may be general anasarca with effusion into the serous sacs.

Heart.—The heart is generally hypertrophied, that is to say, the apex beat is displaced outwards. A systolic murmur is often heard at the apex; when this is absent there is reduplication of the first sound, and the aortic second sound is accentuated.

Pulse.—The pulse is usually about sixty, and incompressible. The sphygmographic tracing (*Fig.* 46,) shews the character of high tension very well, though this particular instance illustrates the statement already made, that the curve of high tension represents only the prolonged systole resulting from the effort of the ventricle to overcome the peripheral obstruction, as it was taken from a case where the heart was failing rapidly.

Fig. 46. High tension pulse tracing from case of chronic febrile nephritis.

Ophthalmoscopic appearances.—According to my experience these are normal in undoubted cases of chronic nephritis belonging to this class. If changes occur they do so very rarely.

Urine.—The *quantity* of urine is generally increased, being from sixty to ninety ounces *per diem*. It is usually *acid*; of a *density* of 1010 to 1015; the *urea* varies from ·7 to 1·5 per cent., the daily quantity being from two hundred to three hundred grains; the albumen is generally considerable in amount, from ·4 to 1 per cent. or more; blood in traces is commonly present; numerous epithelial, fatty, granular, and hyaline casts are visible in the deposit, together with red and white blood corpuscles and fatty renal epithelium.

Headache and *vomiting* are of frequent occurrence. *Convulsions* and *coma* and other symptoms of *uræmia* are less common than in the acute attack, and much less than in the contracting kidney with which they are specially associated, and where they are described fully.

DIAGNOSIS.—When dropsy is present, and the urine is copious, highly albuminous, and deposits numerous casts and renal epithelium, this form of Bright's disease is suggested. The diagnosis may be made quite plain by the history. If this is wanting it may be difficult to decide, as after one or more intercurrent acute attacks the clinical symptoms of lithæmic kidney are so modified as to make the differentiation almost impossible. The presence of retinal changes indicates lithæmic kidney. The age of the patient is of importance, as the latter disease is very common after forty years of age, while primary acute nephritis and its sequelæ are relatively rare.

DURATION.—The disease always lasts months, and may last years. One gentleman well known to me had an attack of scarlatinal nephritis twenty-four years ago. According to his medical attendant his urine always has presented the characters which I have observed in it during the last few years. It contains a large amount of albumen with epithelial, fatty, granular, and hyaline casts. It is possible that the disease in this case affects only one kidney or a portion of one, for such cases have been known to occur. At any rate, it shows that persons presenting the urinary signs of chronic nephritis after an acute attack may go on for half a lifetime. This gentleman is now aged forty; he enjoys fair health, and, although he leads a careful life, he takes ordinary diet, with a moderate amount of wine.

PROGNOSIS.—Although these cases may last a long time, the prognosis must be very guarded, as they are certainly very precarious lives. As a rule they recover sufficiently to go about, but they are very liable to relapse.

The worst prognostic sign is extensive and obstinate dropsy.

ILLUSTRATIVE CASES.

CASE 12*. *Phthisis; chronic febrile nephritis; lardaceous degeneration. Death. Autopsy.*

---

*Recorded by Dr. Stacey Wilson, House Physician.

T. T.——, æt. thirty-seven, gun finisher, admitted April 6th, 1886, complaining of cough, pain in chest, scantiness of urine, and inability to take solid food. He spits a good deal and the sputa have been once or twice streaked with blood.

He has had a cough for eighteen months, and has gradually got weaker, and lost flesh. Six weeks ago diarrhœa set in, and reduced him greatly. He attributes his illness to repeated colds. His work exposes him to dust, but not to cold or damp.

His father, aged sixty-three, is at home suffering from his lungs, but his mother and all his brothers and sisters are alive and healthy.

Present condition.—Patient is a sallow, poorly nourished man; legs and feet œdematous. T. 100°; P. 96; R. 18; tongue clean, red and dry; suffers from flatulence, pain between shoulders, and nausea. Bowels confined. Vertical liver dulness six inches in mammillary line. Spleen not enlarged.

Heart's apex in fifth intercostal space, where there is a faint systolic murmur; first sound reduplicated in tricuspid area; second sound in pulmonary area accentuated and reduplicated.

Pulse small, regular, soft and compressible; voice weak; phonation not painful; cough troublesome; sputa copious, mucous and frothy.

Dulness at both apices, more marked on the right side. Bronchial breathing heard above right clavicle, and whispering pectoriloquy all over right upper lobe. Everywhere else respiration indeterminate, with scanty crepitations.

Urine seventy-six oz.; sp. gr. 1008; acid; bright amber colour; mucous deposit; urea ·6 per cent.; a faint cloud of albumen; hyaline casts and leucocytes in deposit.

A sphygmographic tracing shewed that the tidal

wave was prolonged in spite of the weakness of the heart.

The œdema of the legs at first disappeared, but it returned by April 21st. He got gradually worse. The quantity of urine remained large and the albumen increased in amount.

He died on May 14th of a sudden attack of dyspnœa.

AUTOPSY, May 14th.—*External Appearances.*—General pallor and emaciation; œdema of feet and legs.

Lungs, adherent; honeycombed with cavities, but especially the upper lobe of the right lung; bronchi dilated, containing pus and surrounded by indurated areas. Small patches of grey miliary tubercle scattered through lungs. Both bases congested and œdematous.

Heart, 13 oz.; *left* ventricle dilated and hypertrophied; soft reddish vegetations on aortic valve and on aortic segment of mitral valve.

Peritonæal cavity full of fluid.

Liver, 67 oz.; smooth, tense, edge rounded, colour dull brown; on section, translucent, gave waxy reaction with iodine.

Spleen, 8 oz., a typical "sago spleen."

Kidneys, 9 oz.; capsules partly adherent; surface pale yellowish white, with well defined stellate veins on surface. On section, cortex of normal width; mixed greyish white and yellow in colour; rather anæmic, translucent and shining, giving the iodine reaction distinctly. The mucous membrane of the whole alimentary canal gave the reaction with iodine, but the muscular substance of the heart and tongue was normal.

CASE 13.*—*Chronic febrile nephritis; great improvement.*

---

* Recorded by Dr. Stacey Wilson, House Physician.

Henry M., aged nineteen, warehouseman, admitted March 22nd, 1886, with swelling of the face, feet, and legs. He had been ill since November, 1885, when the swelling began, and he was told at the Dispensary that he had Bright's disease.

He had scarlatina when he was four years old, but no other definite illness.

His father, mother, and six brothers and sisters were all healthy.

On admission he looked very anæmic, with a pale puffy face, and considerable œdema of the feet and legs. T. 98°; P. 72; R. 18; tongue furred; bowels confined; liver and spleen are a little enlarged, the latter can be felt below the ribs.

Lungs, resonance deficient at both apices; cog-wheel respiration; no accompaniments or cough. Heart's apex three-quarters of an inch external to vertical nipple line in fifth intercostal space. Sounds in mitral area very loud. A systolic murmur audible in aortic area, faintly conducted into vessels of neck. Aortic second sound accentuated.

Pulse full, of high tension, artery feels thickened.

Ophthalmoscopic appearances normal.

Urine 60 oz.; sp. gr. 1015; acid; pale yellow colour, smoky; mucous deposit; urea 264 grains *pro die*; albumen ·4 per cent.; blood in quantity; fatty epithelial, hyaline and granular casts, with blood corpuscles and fatty epithelium in the deposit.

He made fair progress; he was kept in bed till May 14th, when the œdema had entirely disappeared. On being allowed up it returned slightly in the legs. He was discharged on July 2nd to go to the Sanatorium, and from there he went to take a situation in the South of England, whence he wrote to say that he was going on very well.

The following table marks his progress:—

Date	Urine in oz.	Sp. gr.	Albumen in grs.	Urea in grs.	Casts
Mar. 24	60	1015	105	264	Epithelial, hyaline, fatty & granular
,, 30	56	1015	81	369	,, ,, ,, ,,
April 8	72	1015	126	285	,, ,, ,, ,,
,, 15	70	1016	308	277	,, ,, ,, ,,
,, 22	72	1015	169	221	,, ,, ,, ,,
,, 27	70	1015	184	215	Hyaline, fatty, and granular
May 7	76	1013	200	299	Epithelial and hyaline
,, 13	72	1012	253	223	Few hyaline and granular
,, 19	80	1014	150	352	Epithelial, fatty and hyaline
,, 25	76	1016	133	334	,, ,, ,,
June 2	84	1014	221	332	Epithelial and hyaline
,, 9	74	1015	162	205	Epithelial and hyaline
,, 16	84	1015	147	221	Epithelial, hyaline, and fatty
,, 23	76	1015	133	367	Hyaline and few granular
,, 30	76	1014	133	334	Hyaline

Treatment.—The patient must be kept in bed, clothed in flannel, in a room the temperature of which is carefully regulated by day and night to avoid chills, and he should be kept in bed so long as any dropsy remains, though slight swelling of the legs coming on after he is allowed up does not necessarily call for a return to bed. The diet should be at first milk, with bread and farinaceous puddings, white fish and poultry being added as the case progresses satisfactorily. Butcher's meat and cheese should be forbidden so long as the doctor has control of the case. No alcohol should be permitted.

The treatment should be chiefly directed to getting rid of the dropsy by diuretics, purgatives, and diaphoresis.

The *hot-air bath* should be used daily. An electuary of bitartrate of potash and honey should be taken freely so as to act upon the bowels, combined with a diuretic pill such as the following:

℞ Pulv. digitalis
   Pulv. scillæ
   Caffeinæ citratis, āā gr. j.
   Ft. pil.
Sig. One to be taken thrice daily.

If the dropsy is very great, speedy relief may be obtained by tapping. If there is much ascites this should be done at once. Tapping the legs by the small trocars introduced by Southey is not a very satisfactory proceeding. The fluid drains away so slowly that the dropsy may not be affected by it. But the great objection is that the punctures are apt to become the seat of erysipelatous inflammation. This practice is sometimes attended with satisfactory results, but it is not one of which I can speak very highly.

When convalescence appears to be established iron should be given, preferably the carbonate, citrate or tartrate.

The precautions to be taken by the convalescent are the same as those already detailed for the acute attack; he should especially guard against acquiring the uric acid dycrasia.

BIBLIOGRAPHY.

AUFRECHT. "Pathologische Mittheilungen," 1 Heft, p. 72.

BAMBERGER. *Op. cit.*

BOUCHARD (C.). Des néphrites infectieuses. "Revue de Méd.," 1881, I., p. 671.

————. Leçons sur les auto-intoxications dans les maladies. Paris, 1887.

CARPENTER (G. A.). Dangers of the continuous hot wet Pack in Acute Renal Disease. "Practitioner," vol. XLI., 1888, p. 178.

COATS (J.). Acute interstitial inflammation of the Kidneys in Scarlet Fever, fatal on the tenth day. "Brit. Med. Jour.," 1874, II., p. 400.

CROOKE (G. F.). Contribution to the pathological anatomy and histology of Scarlatina. "Birm. Med. Review," vol. XXI., 1887, p. 256; vol. XXII., p. 10.

FINLAYSON (J.). The clinical significance of albuminuria; influence of toxic agencies. "Glasgow Med. Jour.," vol. XXI., p. 262.

GREENFIELD (W. S.). A résumé of the present knowledge of Renal Pathology. "New Syd. Soc. Atlas of Pathology," Fasc. II., 1879.

KLEIN (E.). Report on the minute anatomy of Scarlatina. "Reports of Med. Off. Privy Council," N.S., No. VIII., 1876, p. 24.

——————. On the minute anatomy of Scarlatina. "Path Soc. Trans.," vol. XXVIII., 1877, p. 480.

LITTEN (M.). Einige Fälle von mycotischer Nierenerkrankung. "Zeitsch. f. Klin. Med.," Bd. IV., p. 191.

MAURIAC (C.). Syphilose des Reins. "Arch. Gén. de Méd.," 1886, II., pp. 385, 553, 649.

MIRCOLI (S.). Primäre mykotische Nierenentzündungen der Kinder. "Cent. f. d. Med. Wiss.," 1887, p. 738.

SCHELTERNA. Nephritis after Whooping Cough. "Werkbl. van het Nederl. Tijdschr. voor Geneisk.," 1888, p. 166.

SMITH (W. G.). *Op. cit.*

STEWART (GRAINGER). The influence of acute infectious diseases upon the Kidneys and their Functions. (Internat. Med. Cong., Copenhagen, 1884.) "Brit. Med. Jour.," 1884, II., p. 468.

WILLEY (C. H.). A contribution to the pathological anatomy of Scarlatinal Kidney. "Med. Press and Circ.," Dec. 26, 1888.

Chapter XII.
LITHÆMIC NEPHRITIS.*
(Syn. Gouty Kidney.)

This most common form includes a number of cases of acute nephritis, *e.g.*, those occurring in gout, and such instances as that of a habitual beer drinker, with his blood loaded with uric acid, meeting with a severe chill, his kidneys being already more or less irritated or actually altered by the dyscrasia; as well as the very common intercurrent acute attacks which occur during the course of the chronic process. But the very great majority are chronic cases of the clinical type associated with the small red granular kidney, although this limitation to one anatomical type is a mistake.

ETIOLOGY.—As the name implies, it is primarily due to excess of uric acid in the blood and to the prolonged effect of its elimination through the kidneys.

Age.—It is rare under twenty years of age, less rare up to forty, becomes common after that period is passed, and after fifty is so common that nearly one-third of all persons dying above that age show more or less signs of its action in their kidneys.

Sex.—All statistics agree in showing that the contracting type of kidney is less common in females than in males. This is true, but the truth would be more striking if the figures were not vitiated by two circumstances, namely, the frequency of contracting kidney of obstructive origin in women with pelvic diseases, and, secondly, the fact that contracting kidney may be of febrile origin.

Nevertheless, I accept the figures for what they are worth. Dickinson gives the proportion of one female to two males; Wagner, of fifty-five females to ninety-five

---

* The other toxæmic nephritides are of too little clinical importance to call for description.

males. We are not told whether the numbers were calculated out so as to show the proportion to the total deaths in each sex. This is important, as out of a hundred cases collected without reference to etiology from the General Hospital registers, I found thirty females to seventy males, but there were only a hundred and thirty-six female deaths to three hundred and four male deaths, so that the percentages brought the figures very near, namely, twenty-two females to twenty-three males.

But on the other hand, the relative infrequency of this form of Bright's disease in females is an incontestable clinical fact.

Temperament.—The doctrine has been favoured by authority that nervous worry is a cause of contracting kidney, but in my opinion it is the uric acid diathesis which is so often associated with an irritable nervous temperament, the subjects of which are prone to worry.

Social condition.—It attacks all classes, but is especially common among workers in lead, miners, brewers, publicans, forgemen, and stokers.

Heredity.—There should be no doubt of the existence of a hereditary tendency to this disease. In the case already quoted (p. 157) seen with Dr. Lycett, it was this form of Bright's disease from which our patient, his father, and two paternal uncles died.

Eichhorst has related the history of a family, in which the grandmother, mother, two sons and a sister suffered from this disease.

Mahomed believed in the existence of a diathesis—Bright's diathesis—closely allied to gout, and presenting the following characteristics: " Habitual constipation, some forms of dyspepsia, often signs of imperfect circulation, such as cold hands and feet, not unfrequently palpitation, sometimes shortness of breath on exertion. Their skins are often thick, of velvet-like softness and very white." This quotation is given merely to indicate Mahomed's line of thought. So far as my clinical experience goes, I cannot say that it confirms his description of the characteristics of the individuals in whom we commonly meet with this form of Bright's disease, nor am I able to suggest a better one, though I incline to the opinion that the spare neurotic type is especially prone to suffer from lithæmic kidney.

Climate.—The mode in which our moist cold climate predisposes to Bright's disease has been already explained.

Garrod states that chilling the skin increases the formation of uric acid, while it probably gives rise to other alterations in the blood by diminishing elimination.

Gout.—There can be no doubt of the frequent association of these conditions. This was first pointed out by Todd, who gave to the renal disease the apt name of "Gouty Kidney."

Ebstein says that the kidneys may be perfectly healthy even when the articular affection is very pronounced; or they may be in a state of chronic atrophy with uratic deposits; or there may be uratic deposits in the canaliculi or in necrotic foci. He believes the deposit of urate of soda in necrotic foci alone is typical; he recognises the first stage as an inflammatory process set up by uric acid, and followed by necrosis, uratic salts being deposited in the cavities thus formed.

There is no necessary connection between articular gout and nephritis. The tendency to accumulate uric acid in the blood may exist apart from articular gout. It is known that the first metatarso-phalangeal joints of the subjects of this form of Bright's disease are generally the seats of uratic deposits, but a history of attacks of gout is exceptional. Both the joint affection and the kidney disease depend upon the accumulation of uric acid in the blood, but for the production of the former there is needed the presence of special etiological factors.

On the other hand, the kidneys may be quite healthy, in spite of pronounced articular gout. There are people with marked hereditary predisposition who suffer from gout without any indulgence in food or drink. The explanation of this paradox probably is to be found in the excess of uric acid being in these cases a local and not a general condition.

Lead.—Ollivier originally drew attention to the frequent occurrence of the disease among lead workers. He proved that lead is eliminated by the kidney. We know, too, that lead leads to accumulation of uric acid in the system, either by depressing the function of the liver or by forming insoluble urates. There is a general agreement among clinical observers in this country as to its prevalence among patients in whom evidence is to be found of past or present poisoning by lead.

Charcot and Gombault succeeded in causing contracting nephritis in rabbits by the prolonged administration of lead.

Hard water.—Permanent hardness in drinking water, due to excess of lime salts, is an undoubted cause of the uric acid dyscrasia in the persons who use it, and thereby it is an indirect cause of Bright's disease. Its *modus operandi* has not been clearly explained. Whether it acts by forming insoluble urates or by favouring by its presence the decomposition of soluble urates, are questions worthy of investigation. Throughout this district (Warwickshire, Staffordshire and Worcestershire,) the drinking water is very hard, and its relation to uric acid formations is to my mind beyond all question.

Alcohol.—There is no doubt that beer-drinking is a very important cause, but it is probable that beer owes its ill-effects more to the acid than to the alcohol it contains. Dickinson's criticisms of the alcoholic etiology of Bright's disease are entitled to great weight, but it is doubtful whether alcohol can be altogether acquitted.

There is no doubt that alcohol causes cirrhosis of the liver, and a substance that can produce so much organic damage in the great urea-forming organ of the body cannot be innocent of indirectly causing increased formation of uric acid.

Animal Food.—The excessive use of *butcher's meat*, such as is common in this country, especially among certain classes of well-paid artizans, is generally and most properly believed to be a very powerful cause of this disorder. It acts by increasing the raw material from which uric acid is formed, and also, (and this is perhaps more important,) by the large amount of salts contained in it (chlorides, sulphates and phosphates,) which diminish the alkalinity of the blood, and prevent the solution of uric acid.

Dyspepsia.—Johnson has stated his belief in the frequent occurrence of this disease in persons who suffer from certain forms of dyspepsia, or who eat and drink to excess. Murchison was persuaded of its relation to the digestive derangement which he called "lithæmia," and associated with a functional deficiency of the liver. Mahomed published certain cases of young adults in whom dyspepsia was associated with albuminuria and high arterial tension, suggesting the belief that some products of faulty digestion were producing a dyscrasia, which affected the vascular system and irritated the kidneys in such a way as to lead in time to the development of contracting kidney with cardiac hypertrophy.

This is a view which I have fully accepted. It explains the occurrence of this form of Bright's disease in persons who are in no way given to excess, but whose digestive functions are inadequate, either congenitally or as the result of sedentary habits.

Heart Disease.—The relations of heart disease to the contracting granular kidney have been insisted upon by Dickinson, and a study of *post mortem* registers proves it to be a fact. As has been often pointed out the effect of heart disease is to approximate the mammal to the reptile, to diminish all oxidation processes, and to increase the formation of uric acid while diminishing that of urea.

MORBID ANATOMY.—Acute nephritis is not in itself a disease that often terminates fatally. We know it best in the febrile form, where death is due to the primary disease. It is only in very severe cases that we get an opportunity of seeing the kidneys in acute lithæmic nephritis. They are then swollen, of a red or chocolate colour; the capsule strips readily off; on section blood drips from the cut surface, upon which the Malpighian bodies stand out as dark red points.

The condition is one of intense congestion, with acute catarrhal inflammation of the epithelium of the convoluted tubes.

In sub-acute cases the kidneys are still swollen, the cortex increased, pale and mottled; the capsules separate easily, leaving a pale marbled surface.

The changes are similar to those already described in sub-acute nephritis of febrile origin. The epithelium of the convoluted tubes is swollen, granular, vacuolated and fatty; in places only a narrow band is left. The glomeruli are swollen, and their nuclei increased. The blood vessels are full of blood.

Still later, such kidneys may present the appearance of the large white kidney. They are greatly swollen, very pale, or mottled red white and yellow, soft, friable, capsules not adherent; on section the Malpighian bodies appear very distinct.

Under the microscope the epithelium is very fatty; the Malpighian bodies show well-marked glomerulitis; the blood vessels are dilated, and their walls thickened by hypertrophy and *endarteritis obliterans;* the connective tissue is swollen, hyaline and nucleated.

But the typical kidneys of this form of Bright's disease are the small red kidneys. They are small, weighing together less than 8 ounces, hard, with opaque thick adherent capsules, which, when stripped off tear the kidney substance, leaving a dark, red, granulated, often nodular surface, in which are a few small cysts varying in size from a pin's head to a pea. On section, the organ is tough; the cortical portion is dark red dotted with vascular spots and divided from the medullary cones by a well-marked line of hyperæmia; the cortex is very narrow, measuring often from $\frac{1}{8}$th to $\frac{3}{16}$ths of an inch in breadth. The medullary portion is purple, and the cones are striated with lines of hyperæmia. The mouths of the cut vessels are everywhere patent and stiff.

Under the microscope the following changes can be seen:—

Connective tissue.—If a thin section is brushed out so as to remove all but the frame-work of connective tissue, it will be seen that wedge-shaped processes of thickened and nucleated connective tissue pass down into the cortex from the capsule, occupying the region of the interlobular arteries and Malpighian bodies. The intertubular stroma is generally swollen and nucleated.

Malpighian bodies.—The most general change is an increase in the nuclei of the capillary tuft and its conversion into a simple cellular mass. In a smaller number a later stage may be followed, in which the cellular mass becomes converted into delicate gelatinous tissue containing a few stellate cells (*Fig.* 47). Still later the contents undergo a complete colloid change so as to form a small cyst.

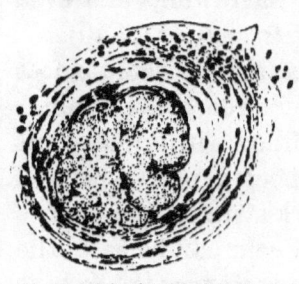

Fig. 47. Malpighian body, shewing well-marked capsulitis, with hyaline degeneration of the glomerular tuft.

Blood vessels.—The capillaries are dilated and full of blood; in places they show concentric hypertrophy (*Fig.* 48). The larger vessels are dilated and their walls thickened. This thickening is most generally seen to affect the adventitia and the muscularis; the intima is thickened frequently, but not so constantly.

Fig. 48. Capillary vessel, shewing concentric hypertrophy of its wall.

In the *adventitia* the change seems to be a purely inflammatory overgrowth which it shares with the neighbouring connective tissue with which it is directly continuous and forms part.

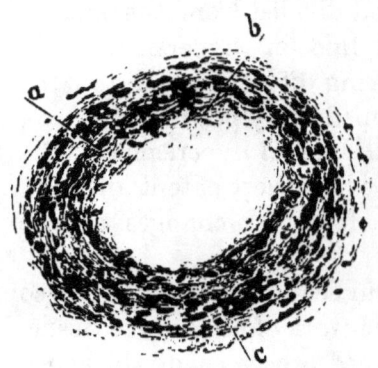

Fig. 49. Renal arteriole, shewing endarteritis obliterans. *a*, Swollen elastic lamina, its fibres separated and œdematous; *b*, Broad growth of connective tissue from the endothelial layer; *c*, Swollen muscular coat, with large distinct nuclei.

The change in the *muscularis* is generally a true muscular hyperplasia, but in some vessels there is dilatation without hypertrophy, as might be expected. In the *intima* the elastic lamina is always swollen, its layers separated, and interspersed with a few nuclei; the endothelial layer is often normal, but sometimes there is a considerable overgrowth of tissue on the inner side of the elastic lamina evidently due to inflammatory thickening of the endothelial layer. Such growths may cause irregular narrowings and even occlusion of vessels (*endarteritis obliterans*) (*Fig.* 49).

Convoluted tubules.—The tubules are in some places dilated, generally they are normal or undergoing diminution in size. They are sometimes filled with amorphous granular matter. In the least affected parts the epithelium is altered; it has lost its dark striated appearance and has become pale, the individual cells can be seen, the nuclei are quite distinct, but in some instances fail to take up staining fluids (*Fig.* 50). But for the most part the tubules are lined by a very flat nucleated epithelium, apparently the atrophied representative of the original cell layer (*Fig.* 52); the cavities of the tubes contain rounded proto-

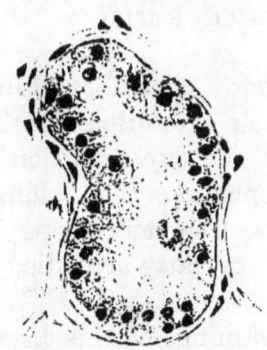

Fig. 50. Convoluted tubule, shewing epithelium with distinct outlines, granular protoplasm and nuclei, which take on carmine staining well.

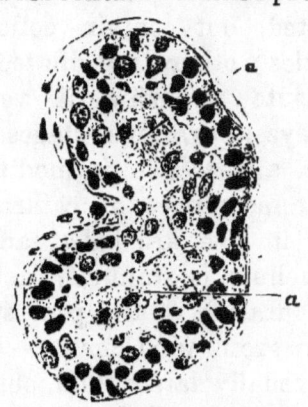

Fig. 51. Convoluted tubule, with proliferated epithelium and casts in lumen.

plasmic masses, nucleated cells, or colloid cast material (*Fig.* 51). Other tubules are denuded of epithelial lining. In other instances the tubules form cysts. This process is effected in a manner very like that already described in the Malpighian bodies. A cellular mass is formed in the tube from the proliferation of the epithelium; this undergoes a retrogressive metamorphosis into gelatinous tissue and thence into simple colloid matter.

The basement membrane of the tubule becomes swollen and hyaline, and is lost in the new formation of connective tissue or becomes the wall of a cyst.

Straight tubules.—The epithelium is often proliferated; the lumina of the tubules are

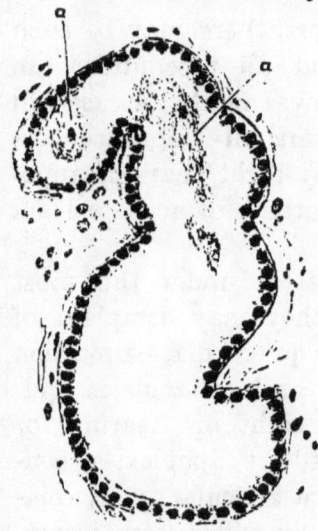

Fig. 52. Dilated convoluted tubule, with atrophied epithelium; *a, a,* fragments of casts.

filled with colloid material mixed with cells. Other tubules are dilated and lined with atrophied epithelium. In the *medulla* many of the straight tubes are unaffected, but contain colloid matter (casts) in their cavities; others are dilated.

Acute nephritis is very prone to attack these kidneys at various stages of their evolution, so that their appearance is modified by the superaddition of the appearances of catarrhal nephritis. The kidneys vary in size, are paler, and under the microscope the epithelium is granular, vacuolated or fatty according to the duration of the process.

SYMPTOMS AND COURSE.—The beginnings of this disease are usually latent and obscure. Its symptoms are in many cases so little noticeable that patients continue to disregard them until some serious and too often fatal complication necessitates medical advice and leads to the discovery of grave organic disease, which has advanced to its full development without any means having been taken to arrest it.

Even after a warning of this sort, there may be such a return of apparent health that all precautions are neglected, until a fresh attack proves fatal. A case will be quoted in which a single uræmic fit diagnosed correctly at the time was followed by eight years of apparent good health until a sudden attack of acute œdema of the lung proved fatal.

Such patients present themselves under the most varied semeiological conditions; they may complain of bronchitis or asthma, pain at the præcordia, palpitation or epistaxis, vomiting, diarrhœa, or hæmatemesis, giddiness, headache, affections of sight or hearing, or neuralgia; they may be attacked by apoplexies, convulsions, or coma; they may have articular gout, sciatica, lumbago or cramps; they may suffer from hæmaturia or symptoms of gravel or calculus; or they may chiefly complain of a skin affection—pruritus, eczema, erythema, purpura, etc.

Fortunately it is rare before forty years of age, but in patients after that time of life we must be for ever alive to the possibility of this condition underlying the complaint put forward. We can only escape falling into mistakes in practice by care, and the importance of making urinary examinations cannot be too strongly insisted upon.

Very frequently the patient comes first under observation owing to the supervention of an attack of acute catarrhal nephritis, by which the whole aspect of the case is temporarily altered.

The urine.—The characters of the urinary secretion are naturally the most important, unequivocal and significant of the symptoms of all forms of Bright's disease, nor are they less so in the present form, though the departures from the normal may not attract the patient's attention.

It is impossible to state precisely at what period in this insidious malady the urine shows signs of alteration, but it may be affirmed with certainty that no marked structural alterations can take place without evidence of their being discoverable in this secretion.

The *quantity* of urine, as in other forms of Bright's disease, undergoes alteration; but in this form it is not diminished in the early stages. On the contrary, when these cases come under observation the quantity is generally above the average, sometimes reaching or exceeding a hundred oz. in twenty-four hours. But this increase, *as indicated by the necessity to rise at night to pass water,* is not always an early symptom. William S—— (Case 15, p. 216), who died six weeks after admission with advanced contracting kidneys, stated that he had only been disturbed at night for a week past. So, too, in the case of Cornelius H—— (Case 20, p. 228,) this symptom had existed only three months, though he had all the other symptoms of an advanced case. On the other hand, Richard J—— (Case 27, p. 254,) had been in the habit of getting up in the night to make water five or six times for the last ten years. But in all these cases its occurrence coincided with the development of the first symptoms which attracted the patient's attention. In a series of cases tabulated some years ago, the necessity to rise at night to pass water was present in seventy out of a hundred cases able to attend as out-patients.

It must be remembered that occasional rising at night for this purpose is common as a consequence of too free potations, of dyspepsia and of excessive tobacco-smoking; while in local diseases of the bladder and neighbouring organs, it is often a marked symptom.

In the later stages the quantity falls; this is due to failure of the heart, and even before irrecoverable collapse of the heart, the urine is liable to be diminished by its temporary weakness. In the case of George B—— (Case 25, p. 247,) it is recorded that "he used to get up twice at night to make water and filled the vessel three parts full, but *since his illness this had passed off.*" During the occurrence of an intercurrent attack of acute catarrhal nephritis the urine is diminished, and presents the usual character met with in primary acute attacks.

The *density* varies with the quantity, and particular samples passed during the day may be very much higher than the total quantity. In general it is below 1010. The *colour* is usually pale yellow, and the urine is *clear*, depositing only a very slight mucous cloud. The *reaction* is almost invariably acid, except when the patient is taking alkalies.

Bartels and Grainger Stewart state that the *urea* may not be diminished. This is true, as, for example, in the case of William S—— (Case 15), who, after living on milk diet for twenty days, passed 343 grains of urea in twenty-four hours. I have not felt justified in endeavouring to ascertain what maximum amount could be eliminated by these patients on full diet. As a rule, however, they pass too little urea, that is, the daily output is under 300 grains.

The excretion of *uric acid* is stated by Frerichs to be lessened, especially towards the termination of the disease.

The *phosphates* are notably diminished; the *chlorides* are diminished, but become normal or in excess (LÉPINE) towards the end; the quantity of *sulphates* varies greatly, but is ordinarily low.

The *deposit*, as already stated, is scanty; it contains a few small hyaline and granular casts, but although not numerous they are very constantly present, and will be found if looked for in the proper way. In acute attacks the deposit is characteristic of acute nephritis.

Blood is not commonly present apart from acute attacks, but hæmaturia may occur, and is sometimes so profuse as to cause death from hæmorrhage (WEST).

Albumen is generally to be observed, if it is properly looked for. It may be absent in the night, but is present in the urine passed during the day, especially after breakfast. The explanation of the greater frequency with which I met with albumen amongst my out-patients was that I saw them in the morning, whereas in many places, notably in London, out-patients are seen in the afternoon; also that I always made my patients pass water for me at the time, and did not encourage them to bring it in bottles, as such urine is generally that passed the first thing on rising and has been secreted during the night.

Bartels has recorded a case which was kept under close observation until its termination, without the presence of albumen having been at any time detected.

The best method of testing for albumen is by boiling

and acidulating with dilute acetic acid, which has been already fully described.

The quantity of albumen is small in uncomplicated cases, but rises as the urine diminishes, following the law already established that albuminuria is increased by lowering the blood pressure. It is also increased in intercurrent attacks of acute nephritis.

Saliva.—This secretion has been found to contain urea (FLEISCHER) and albumen (SEMMOLA, VULPIAN and STRAUSS) in certain cases.

The blood.—Anæmia is a marked symptom. Leichtenstern found the hæmoglobin co-efficient reduced from the normal 1330 to 802; Dickinson found the red corpuscles reduced from the normal 5,000,000 to 3,921,875, and Rosenstein to 3,000,000. The water of the blood is increased from 784 parts *per mille* (CHRISTISON, OWEN REES, RAYER) to 821-853 parts *per mille;* while the albumen is diminished from 73·4 *per mille* to 68·5-59 *per mille* (OWEN REES, and RAYER). The urea is increased from the healthy standard of 0·016 to 0·084.

The heart. — The changes in the heart found *post mortem* in their order of frequency are: (1,) Hypertrophy, which is present in about 60 per cent.; (2,) Atheroma of the aorta, coronary arteries or endocardium, leading in the latter case to thickening of the valves without definitely impairing their functions; (3,) Valvular disease affecting in about equal proportions the aortic and mitral valves, stenosis of the latter valve being apparently much more common than simple dilatation; although this may be because it is so much more definite that the reporter has less difficulty in stating the fact; (4,) Fatty degeneration (granular atrophy) of the muscular fibre of the wall of the heart; (5,) Pericardial adhesions; (6,) Pericarditis. Pericardial effusion as part of general dropsy is not uncommon, but less so than pleural effusion.

The only definite sign of cardiac hypertrophy is displacement of the apex beat. This is normally in the fifth left intercostal space, well to the inner side of a line drawn vertically through the nipple. From my own observations, made on numerous out-patients, I am unable to accept the rigid descriptions that limit the normal position more than this. If the apex beat is in the nipple line or to the left of it, the heart is enlarged; still more manifestly is this the case if the impulse is in the sixth interspace, instead of the fifth. When there is much hypertrophy the impulse of the heart is often strong and diffused over a large area.

The first sound of the heart at the apex is commonly reduplicated, while the second sound in the aortic area is accentuated.

Johnson has suggested that the doubling of the first sound is due to the contraction of a dilated and hypertrophied auricle becoming audible, but the more generally accepted explanation is that of Sibson, who ascribed the reduplication to the asynchronous action of the two ventricles, due to the greater difficulty the left ventricle has in discharging its contents owing to the increased pressure in the aortic system; while he explained the unity of the accentuated second sound by suggesting that the increased tension in the aorta allows it to complete the closure of its valves synchronously with the earlier filled but less actively distended pulmonary artery.

These changes in the heart sounds are very constant in contracting kidney, but their diagnostic value must not be over-estimated.

Doubling of the first sound may be heard in bronchitis and emphysema when there is obstruction to the discharge of the right ventricle, and in mitral constriction in cases where no murmur may be audible.

Accentuation of the aortic second sound is common in

youths whose hearts under examination act with more than wonted energy. It may also be present wherever any local cause in the thorax (*e.g.*, tumour, aneurism) raises the blood pressure in the aorta.

But these changes in the heart sounds are so readily ascertained that they are of great value, as they often indicate the necessity for further careful investigation, and by attracting attention lead to the recognition of the renal condition which otherwise might pass unobserved.

Murmurs are not uncommon in contracting kidney. Systolic mitral murmurs may be due to dilatation or to the accidental association of old rheumatic endocarditis. According to Bartels acute endocarditis may occur as a result of renal disease. Aortic murmurs of obstruction or regurgitation, systolic and diastolic, are generally due to chronic endarteritis deformans attacking the aorta and spreading to the valves, but of course they sometimes have a rheumatic origin.

The presence of murmurs is an unfavourable element, as the heart has a very hard task to perform to compensate for the renal defect, and if handicapped by a valvular insufficiency it will probably fail early in the struggle.

From the point of view of the prognosis of *heart* disease, the supervention of kidney mischief is very unfavourable. A valvular defect, which was apparently compensated, will acquire fresh importance, and be followed by rapid heart failure and death.

Palpitation is a symptom commonly complained of. It is probably in many cases toxæmic in its origin.

The pulse.—The pulse rate in contracting kidney is generally high, from 90 to 100, but it may be normal, 70 to 80, or low, 60. In character it is usually hard, and incompressible, but varies in size, being sometimes full, more usually small. The hard radial artery, resembling the spermatic cord to the feel, is the typical high tension pulse of contracting kidney. But it is perhaps more commonly absent than present. When the artery is neither hard nor prominent the pulse will still be found to be incompressible, and even when this character is lost, the sphygmographic tracing may still show signs of peripheral obstruction to the circulation (*Fig.* 53 illustrates a typical pulse tracing from contracting kidney).

Fig. 53. Pulse tracing from case of R. J. Case 27). Lithæmic kidney, chronic uræmia.

Ophthalmoscopic changes.—It is especially (if we exclude the nephritis of pregnancy the statement may be made more absolute), in this form of Bright's disease that we meet with affections of the retina. These have already been fully described (Chap. VII); they consist for the most part of hæmorrhages, inflammatory exudations in and around the disc, and degenerative patches chiefly in the neighbourhood of the yellow spot. Sudden loss of vision may be due to hæmorrhage into the yellow spot.

CASE 14.—Frederick J——, aged sixty-three, attended as an out-patient on March 26th, 1881, with pain in bowels, dyspnœa, cough and palpitation. He had been ill three months. He was in the habit of getting out of bed four or five times nightly to pass water. Heart irregular and intermittent. Urine albuminous. He returned on April 9th with loss of vision in the left eye and dimness of the sight of the right eye. Ophthalmoscopic examination showed: *Right eye*, $H = \frac{1}{3}$, disc very vascular; no exudations. *Left eye*, disc hazy, arteries very small, reduced almost to threads. Several large diffuse hæmorrhages *involved the whole region of the yellow spot.*

The following case is a typical example of albuminuric retinitis, and illustrates not only the various forms of retinal lesion, but the changes which may take place in them. As already explained, there is nothing essentially incurable in the retinal disease; its gravity depends upon its relation to advanced and incurable organic disease of the kidneys.

CASE 15*.—*Chronic lithæmic nephritis; albuminuric retinitis. Death. Autopsy.*

William S——, aged twenty-three, brass caster, admitted February 24th, 1888, complaining of headache, pain in the loins and dimness of sight. He had been subject to headache for ten years, had often been giddy, but had not noticed any affection of his eyesight till a week ago, and for the last four days he had been vomiting. He had never had a fit, or been dropsical. He was a teetotaller, his work did not involve the use of lead, and he could remember no illness. His father died of dropsy and one sister of phthisis; his family history was otherwise unimportant. He had lately noticed that his eyelids were swollen in the morning and that his ankles had been œdematous on two occasions; he had been obliged to rise at night to pass water only for the last week. He was a well made man, with a yellow, waxy complexion, and some œdema of the lower eyelids. T. 98·4°. P. 90. R. 20. Tongue furred; bowels confined; liver and spleen normal. Lungs normal, with the exception of a few catarrhal sounds. Heart's apex in nipple line and fifth interspace; at apex reduplication of first sound; a systolic murmur in aortic area. Pulse incompressible, regular. Urine 30 oz.; sp. gr. 1010; acid, albuminous; urea 132 grains; granular, epithelial and hyaline casts, red and white blood corpuscles and renal epithelium in deposit. Ophthalmoscopic examination of right eye showed papillitis, with numerous radiating hæmorrhages along the course of the blood vessels; left eye, papillitis with several large round soft-edged white patches in the neighbourhood.

---

* Recorded by Mr. T. O. Crump, Clinical clerk. This case has been partly described at p. 83.

LITHÆMIC NEPHRITIS.

He went on pretty well for some time. On March 26th the eyes were examined and the right shewed still swelling of disc with absorption of hæmorrhages; the

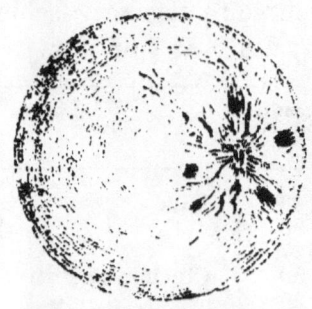 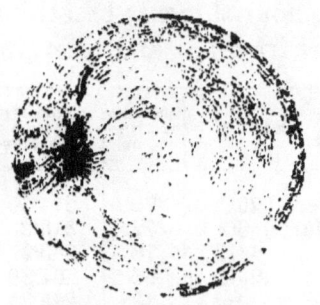

Fig. 54. Right Eye.—Disc swollen and infiltrated, with flame-shaped hæmorrhages radiating round it.

Left Eye.—Disc much swollen and infiltrated, with soft round patches of inflammatory exudation round it.

left eye showed the outline of the disc well defined, the former rounded patches gone, but a fresh set of oval radiating bright patches on the inner side of the *macula lutea*.

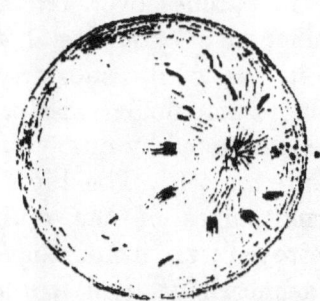 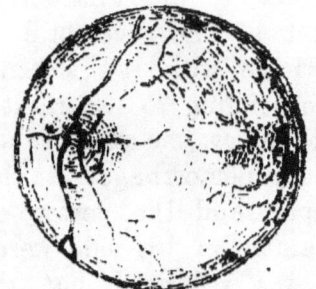

Fig. 55. Right Eye.—Swelling of disc undiminished. Hæmorrhages undergoing absorption.

Left Eye.—Disc much less swollen; vessels visible; rounded patches gone. Patches of retinal degeneration radiating round the yellow spot.

On April 2nd he began to have headache. Temp. 99° F. He was put back on milk diet.

3rd.—Headache continued; partial unconsciousness; passed urine in bed; pupils sluggish to light. Temp. 98·4°.

4th.—Conscious. Temp. 98°.

5th.—Temp. 100° F. Face flushed; tongue dry; headache.

6th.—Temp. M. 100·8. V. 102·5.

7th.—Temp. 101. Unconscious; could not be got to take nourishment; breathing laboured; pulse very feeble; heart irregular and intermittent; death at 7.35 p.m.

Date	Quantity of urine in oz.	Quantity of urea in grains	Quantity of albumen in grains	Diet
Feb. 26th...	78	274·56	63	Milk
March 6th...	72	316·8	60	,,
,, 16th...	78	343·2	117	,,
,, 24th...	66	261·36	60	Chicken or Fish
,, 31st...	73	224·84	120	,,

No urine collected after this date.

AUTOPSY, September 4th, 1888. — Features pallid, nutrition visibly impaired; no œdema of legs, feet or scrotum.

Brain 47 oz.; hæmorrhage in patches over lateral aspect of right cerebral hemisphere; extensive hæmorrhage with red softening in the white substance corresponding to the right temporo-sphenoidal and occipital lobes, the softened area surrounded by numerous miliary hæmorrhages into the white matter. The hippocampus and the postero-external edges of the optic thalamus on this side were destroyed; no hæmorrhage into the ventricle, but the hæmorrhage had oozed through the occipital lobe on to its outer aspect and spread over the whole of the cortex on the right side; serous effusions into the ventricles, the posterior part of the external capsule was involved; arteries at base of brain appeared normal.

Heart 15 oz.; no valvular lesions; coronary arteries good; great hypertrophy of left ventricle; wall measured $\frac{8}{10}$ in. thick. *Muscle* somewhat pale, but looked fairly normal. *Lungs* congested and œdematous, especially the inferior lobes. *Kidneys* barely 7 oz.; contracted, capsule adherent, surface granulated, but also somewhat hyperæmic; on section, the atrophy was more or less uniform; cortex of a mixed reddish grey and yellowish grey colour, and offered perceptible resistance to the knife edge.

Dropsy occurs in the acute forms, and has then the ordinary characters of the dropsy of acute nephritis. But in the commoner chronic form it only supervenes in the later stages as a consequence of heart failure, and then presents the usual character of cardiac dropsy, beginning in the depending parts, and gradually rising as the water gains on the pumping power of the heart.

DIAGNOSIS.—The latent nature of this disease has been referred to many times. It comes under so many guises that it is only by possessing the conviction of its great frequency in persons over forty years of age that the physician will avoid overlooking many cases.

The hard pulse is one of the most accessible of signs, and is fairly constant, being present in 62 per cent. of my cases. Accentuation of the aortic second sound was present in 80 per cent.

Hypertrophy of the heart, indicated by displacement of the apex beat to the left of the vertical nipple line, was present in 60 per cent.; while a history of rising at night to make water was obtained in 70 per cent.

None of these signs taken singly is diagnostic, but they serve as guide posts to point out the way and lead up to the examination of the urine.

Albuminuria was present in 91 per cent. of my examinations, and will be very rarely found absent if the precaution be taken of obtaining a sample of the after breakfast urine. It is worth remembering that a sample of the total mixed urine of the twenty-four hours, so desirable where quantitative analysis is required, is not the best for discovering a faint trace of albumen, owing to the greater portion of the urine, namely, that secreted at night, being often non-albuminous.

Casts are very constant and of great importance. I have found them in 88 per cent. of all examinations, but I do not think I have ever diagnosed contracting kidney until I had found casts, or ever failed to find them some time or another in every one of the cases of this disease which I have recognised during life.

The occurrence of retinal changes is too infrequent to be of great diagnostic value. Retinitis and neuro-retinitis appear to occur in only about 5 per cent., and small specks and hæmorrhages in 25 per cent.

Apart from their inconstancy they are not absolutely diagnostic when present. Typical albuminuric retinitis may be met with in anæmia; and diffuse neuro-retinitis, exactly like that of Bright's disease, has been seen associated with cerebral tumour. The specks are often extremely small and few in number, and are met with apart from Bright's disease. But combined with the other signs enumerated, these ophthalmoscopic appearances have great weight. In making a diagnosis all the symptoms should be passed in review.

Is it possible to differentiate between the two forms of chronic Bright's disease?—It is certain that, except in the case of typical gouty kidneys, it is difficult, if not impossible, to determine the etiology of the kidneys of chronic Bright's disease by their anatomical conditions; it is therefore probable that the clinical diagnosis is not easy.

The following considerations may guide us: (1,) The age of the patient; the primary chronic form being very common after forty years of age. (2,) The presence of certain symptoms, such as polyuria, before the occurrence of an acute attack, although this might be represented as the commencement of the illness.

(3), The absence of dropsy or its presence only in depending parts. (4,) The presence of retinal changes.

PROGNOSIS.—There can be no doubt that this is an essentially incurable disease, and its course is liable to be interrupted by so many serious complications that the duration of life is always uncertain. Nevertheless, its normal evolution is slow.

Bright was aware that some cases last many years, and that this is possible even under unfavourable external conditions the following brief record shows.

CASE 16.—John P——, aged forty-eight, cabman, was in Hospital in 1877, with albuminuria and dropsy, under the late Dr. Russell. I examined his urine several times and found it to contain numerous epithelial, hyaline and granular casts. This was probably an intercurrent acute attack. He attended for some time as an out-patient, and was then lost sight of. July 14th, 1881, he brought his son to see me, and I then took the following notes. J. P. gets up once every night to pass water. Urine pale, clear, 1010, no albumen or casts. Pulse distinct, but not hard. Aortic second sound accentuated. The disease seemed practically cured, but was only latent. On June 14th, 1884, he came up with diarrhœa, looking very ill. His urine was clear, pale, 1010, contained a good trace of albumen, a few granular casts and blood corpuscles. Heart's apex not displaced; aortic second sound not accentuated. February 27th, 1887, gets up two or three times at night to pass water. Urine pale, clear, 1001, a faint haze of albumen. Has athetosis affecting the right hand and forearm.

In the month of February, 1888, I instituted an inquiry into the present condition of a few persons presenting symptoms of contracting kidney, whom I had rejected for life assurance some years back.

1.—T. B. D., male, aged thirty-eight; examined

December, 1880; urine contained a trace of albumen with hyaline and epithelial casts. Reported to be *quite well*.

2.—R. T., male, aged forty-one; examined February, 1881; urine contained a trace of albumen, hyaline and granular casts. Reported to be *still alive and well*.

3.—J. H., male, aged fifty-eight, examined September, 1881; urine contained a faint trace of albumen, and a few hyaline casts. *Could not be traced.*

4.—C. R. G., male, aged sixty, examined November, 1882; urine contained a faint trace of albumen, but no casts. Died rather suddenly in the autumn of 1886.

The number is small, but it is noteworthy that there was only one death, and that that was a man at least sixty-five years of age when he died.

The most unfavourable symptom is *heart failure*, indicated by dropsy of the lower extremities, and diminution of the amount of urine. The presence of valvular disease, or any cardiac complication, is also most grave, on account of the dependence of the case upon the power of the heart to compensate for the renal defect.

Retinal disease is very unfavourable, because it rarely, if ever, occurs except in advanced cases.

Any of the symptoms of *uræmia* are bad, but life may be prolonged after many of them.

Dyspnœa is one of the most fatal of all the uræmic phenomena.

Acute *œdema of the lung* is always fatal.

The gravity of *acute intercurrent attacks* must be estimated by the amount of dropsy, just as in ordinary acute nephritis. The presence of cardiac complications is of course most unfavourable.

All *acute inflammatory* complications are very serious, especially pneumonia, pericarditis and cellulitis, which are probably always fatal.

Any serious accident such as a *fracture* or condition requiring *surgical interference* usually terminates in death. Surgical statistics, especially of herniotomies in elderly people, are greatly prejudiced by this disease.

ILLUSTRATIVE CASES.

CASE 17.—*Sub-acute lithæmic nephritis in a case of heart disease. Death. Autopsy.*

F. I.——, male, forty-one, labourer, admitted April 8th, 1886, with shortness of breath. He had been quite well until Christmas, when he left off work owing to shortness of breath and cough. He attributes his illness to catching one cold after another. His urine has always been natural, and there has been no dropsy. He has been doing very heavy work and exposed to all kinds of weather.

When he was sixteen or seventeen years of age he had acute rheumatism, but can remember no other illness.

Father died suddenly of apoplexy. Mother died of bleeding from the nose. One brother died of a growth in the mouth (? epithelioma).

On admission his legs were swollen, but this disappeared rapidly after he was sent to bed. Tongue pale, moist, slightly furred; appetite bad; bowels regular. Liver dulness commences at sixth rib, and the edge can be felt two and a half inches below the costal border.

Spleen normal. Heart's apex in sixth interspace; a loud double murmur heard best in aortic area and conducted into the vessels of the neck.

Pulse collapsing, 108. Lungs resonant, breath sounds harsh and accompanied by numerous moist sounds.

Urine thirty-six oz., 1027; acid; deep yellow, turbid from urates; urea 2·2 per cent.; a faint cloud of albumen; hyaline casts; leucocytes and oxalates in deposit.

Progress of case.—He improved at first, but towards the end of April his cough became more troublesome. On May 2nd he had a severe attack of dyspnœa, his pulse became very weak, and he died at 1.50 a.m. on the following morning.

AUTOPSY, May 4th. — Well-made muscular subject; œdema of feet and ankles; slight general œdema of subcutaneous tissue.

Lungs very œdematous; about eight oz. of serum in each pleural cavity.

Heart, twenty-one oz.; right ventricle dilated and full of clot; pulmonary and tricuspid valves normal; left ventricle hypertrophied, walls thickened, cavity dilated, containing a small quantity of dark fluid blood; anterior cusp of aortic valve thickened and retracted; base of aorta irregularly dilated and the seat of extensive *endarteritis deformans;* mitral valve normal; muscular wall of heart pale and marbled; a patch of pericarditis on apex of heart with adherent organised lymph.

Liver, fifty-six oz., a good example of cyanotic induration with atrophy; marked fatty infiltration in portal zones of lobules; consistence firm, rather leathery.

Kidneys, eleven and a half oz., cortex increased, pale, swollen and translucent in places, with brick red and yellowish mottling; capsules separated easily, leaving a pale marbled surface (red and greyish yellow).

On microscopical examination the convoluted tubes were generally dilated, in some the epithelium was swollen, granular, vacuolated and fatty; in others it was atrophied or reduced to a narrow border of protoplasm. Except where degenerative changes were present the nuclei stained well. The glomeruli were hypertrophied, the capillary loops appeared cloudy and indistinct and shewed an increase of nuclei. They were not congested as a rule. Here and there small stellar patches of interstitial change were met with. In the pyramids the capillaries were considerably engorged.

CASE 18.—*Chronic lithæmic nephritis; chronic endocarditis; heart failure. Death. Autopsy.*

Thomas S——, aged forty, gardener, admitted March 28th, 1886, with cough, tightness of the chest, weakness and swelling of the legs.

A month ago he noticed his eyelids and hands were puffy, and he made very little water; then his legs swelled; and a fortnight ago he took to his bed.

He could remember no previous serious illness; he had never had dropsy or any trouble with his water before; he had never had gout, or acute rheumatism, but had often had rheumatic pains in his knees and shoulders, and had been confined to the house for a fortnight with them. For the last six months he had been rising at night to pass water.

His home and circumstances were comfortable, but his work exposed him to cold and damp, and he was in the habit of drinking four or five pints of "sweet ale" daily.

His family history was unimportant; father died aged seventy-two; mother died aged seventy-six. Seven brothers and sisters alive and well; only one had died in infancy.

He was a well-built and well-nourished man, with a ruddy complexion; there was œdema of the conjunctivæ, lower extremities and scrotum. T. 97·5; P. 90; R. 20; tongue furred in centre, red at edges; bad taste in mouth; appetite good; very thirsty; bowels confined; abdomen distended; some ascites; hepatic and splenic dulness normal.

Heart's apex in fifth interspace, inside the vertical nipple line; no murmur; no accentuation of aortic second sound. Pulse regular, not hard.

Breath sounds harsh, with moist râles posteriorly. Vision good; ophthalmoscopic appearances normal.

Urine 28 oz.; sp. gr. 1022; acid; urea 217 grains *pro die*; albumen 71 grains *pro die*; epithelial, fatty, hyaline and granular casts, blood corpuscles and blood casts, with renal epithelium in the deposit.

He did not improve under treatment; he had profuse diarrhœa, and on April 7th he was in a very serious condition. In the evening he complained of feeling very weak; he became cyanosed, and died at 9.20 a.m.

AUTOPSY, April 9th, 1888.—Extensive general anasarca; head and face congested; *rigor mortis* passing off.

Thorax. — About a pint of clear fluid in each pleural cavity. *Lungs.* — Congested and œdematous. *Heart* weighed 17 oz.; pericardial sac contained about 2 oz. of fluid; right side of heart full of clot; tricuspid and pulmonary valves healthy; mitral cusps thickened, shortened, and yellow, with small, firmly adherent vegetations on their auricular surfaces; aortic valve incompetent; segments glued together, thickened, stiff and calcareous in places. Under the microscope the heart fibres shewed a little " brown atrophy," but were not fatty.

Abdomen.—*Liver* 79 oz., fatty. *Spleen* 6 oz., soft, pale. *Kidneys* together weighed 17 oz., greatly swollen, mottled red, white and yellow (roan). On section, the cortex was of the same colour; the Malpighian bodies were very distinct, and some showed the waxy reaction with iodine. The consistence of the organs was soft and friable.

Under the microscope the following changes were noted:—*Malpighian bodies;* the capsules were broadened and nucleated (*pericapsulitis*), and the lining epithelium proliferated (*endocapsulitis*); the glomerular tufts were covered with nuclei, and one had completely undergone hyaline change. *Blood vessels* dilated; there was hypertrophy of the muscular wall, with proliferation of the endothelium of the intima (*endarteritis obliterans*). *Connective tissue;* this was generally broadened out and full of round brightly stained cells. The great exception was the capsule of the kidney, which did not appear thickened. *Epithelium;* the tubes were everywhere lined with fatty epithelium; it was difficult to find one that was not so, even in the medulla. There was no appearance of proliferation of epithelium, though some tubes were filled with *débris*, in which some small rounded nucleus-like bodies could be seen.

CASE 19. — *Acute attack supervening in the course of chronic lithæmic nephritis. Recovery.*

Michael H——, aged thirty, iron worker, admitted April 4th, 1888, with headache, tightness of chest, and swelling of face and legs. He had scarlatina when a child, but enjoyed good health till eighteen months ago; since then he had suffered from severe headache and bronchitis. At Christmas he caught cold, and his legs swelled, but this passed off. Three weeks ago he noticed that his water was smoky. He admitted having been a heavy drinker, and his work exposed him to severe changes of temperature. His mother died of asthma. He was a fairly well nourished man, with a flushed face; when examined the œdema of the legs had passed off from rest in bed. T. 98·5; P. 72; R. 83; tongue clean; appetite good; bowels regular; liver dulness three and a half inches in the vertical nipple line; some dulness in flanks, which changed with his position.

Heart's apex in sixth intercostal space, one and a half inches to the left of the nipple line. At apex, first sound reduplicated; in aortic area, second sound accentuated. Pulse was hard and incompressible. Lungs normal.

Urine 56 oz.; sp. gr. 1017; acid; straw colour, smoky; white deposit; urea 344 grains; albumen 377 grains; deposit contained a few epithelial and numerous hyaline and granular casts, with red and white blood corpuscles and renal epithelium.

With rest in bed and diaphoresis he improved rapidly.

April 14th. — There was no trace of ascites. Urine 60 oz.; sp. gr. 1022.

18th. — Urine 62 oz.; sp. gr. 1010; very little blood; albumen 528 grains; granular and hyaline casts with blood corpuscles and renal epithelium.

There was continued improvement.

May 1st.—Urine 60 oz.; sp. gr. 1006; straw colour; albumen 396 grains.

3rd.—Allowed to get up. Urine 62 oz.; sp. gr. 1020; albumen 327 grains.

8th.—Urine 60 oz.; sp. gr. 1015; straw colour; urea 448 grains; albumen 316 grains; a trace of blood with guaiacum and ozonic ether; a few hyaline and granular casts, blood corpuscles and renal epithelium in the slight white deposit.

10th.—He was made an out-patient.

After April 11th he had milk diet with chicken or fish.

CASE 20*.—*Chronic lithæmic nephritis; albuminuric retinitis; uræmic dyspnœa.*

Cornelius H——, aged fifty-eight, admitted on June 12th, 1888, with severe frontal headache made worse by lying down, cough and dyspnœa, frequent micturition especially at night, partial loss of sight and swelling of the feet towards night.

He had been ill six weeks, and worse for seven days. He was sent to me from the Eye Hospital, where he had complained of headache and dimness of sight, and was suffering from nasal catarrh with congestion of the conjunctivæ. Albuminuric retinitis was recognised there, so he was transferred to me.

His father died of bronchitis, aged fifty-six; mother died aged seventy, cause unknown. Three sisters died between thirty and forty, cause unknown. He has had three children, of whom two died in infancy; the third is living, aged twenty-five, and is strong and well.

Patient had worked as a match maker all his life. He denied having been intemperate; his only drink had been half a pint of beer at supper three or four times a week. He had a good home, but his workshop

---

* Recorded by Mr. F. H. Noott, Clinical Clerk.

was close. He could remember no illness, except "congestion of the kidney" eight years before. On inquiry about the congestion of the kidney, he said that he passed blood in his water and was delirious, and that he was ill in bed four weeks. There had been no return of hæmaturia. He denied ever having had gout.

He was a very old-looking man for his years, with a very puckered, wrinkled face, scanty hair; skin of face and conjunctivæ slightly jaundiced; expression of face dull and apathetic. There was slight œdema of the legs and feet. Temp. 98.

Lips cyanosed; teeth very defective; tongue clean; mouth very dry; appetite good; great thirst; bowels confined; no ascites; liver dulness four inches in vertical mammillary line. Heart's apex a little external to the vertical mammillary line in the fifth left interspace; sounds pure; aortic second sound accentuated. Pulse full, cord-like, incompressible, 96.

Respirations hurried, 48; a little cough; no expectoration; breath sounds and percussion note normal. There is persistent dyspnœa.

Ophthalmoscopic appearances.—*Right eye.*—One or two small hæmorrhages below the disc, and a number of small round patches of atrophic retinitis on the outer side. *Left eye.*—No hæmorrhages, but a larger group of the same patches on outer side of disc.

Urine.—Slight uneasiness in loins; water passed six or seven times during the night for three months past; 66 oz.; sp. gr. 1010; pale, acid; a dense cloud of albumen; urea ·9 per cent. (260 grains); under microscope numerous leucocytes and a few hyaline and granular casts.

He complained of being unable to sleep in the hospital, and insisted on going out five days after admission.

BIBLIOGRAPHY.

CHARCOT and GOMBAULT. Note relative à l'étude anatomique de la Néphrite Saturnine expérimentale. "Arch. de Phys.," 1881, p. 124.

DICKINSON (W. H.). On the Pathology and Treatment of Albuminuria. 2nd edition. London, 1877.

EBSTEIN (W.). La Goutte, sa nature et son traitement. Traduction du Dr. E. Chambard. Paris, 1887.

EICHHORST (H.). Handbuch der speciellen Pathologie und Therapie. 1884, Bd. II., p. 62.

FLEISCHER (R.). Quoted by Rosenstein, *op. cit.*, p. 211.

JOHNSON (G.). Lumleian Lectures on the Muscular Arterioles. London, 1877.

LEICHTENSTERN (O.). Untersuchungen über den Hæmoglobingehalt des Blutes in gesunden und kranken Zuständen. Leipzig, 1878.

MAHOMED (F. A.). Some of the clinical aspects of Bright's disease. "Guy's Hosp. Reports," 3rd S., Vol. XXIV., 1878, p. 363.

—————— Chronic Bright's disease and its essential symptoms. "Lancet," 1879, I., p. 46.

OLLIVIER (A). Essai sur les Albuminuries produites par l'élimination des substances toxiques. Paris, 1863.

TODD (R. B.). Clinical Lectures, 2nd edition, London, 1861.

WEST (S.). On the Occurrence of Blood in the Urine in Granular Kidney. "Lancet," 1885, II., p. 104.

Chapter XIII.
OBSTRUCTIVE NEPHRITIS.
(Syn. Surgical Kidney; Puerperal Kidney; Ascending Nephritis.)

INFLAMMATION of the kidney which arises in the course of many affections of the pelvic organs has close pathological relationships with what is generally known as Bright's disease. A description of it will make clearer the general doctrine of the broad unity in type of the anatomical changes met with in all cases of nephritis, and it will add an additional argument to the proof of the efficiency of the renal lesion as a cause of the cardio-vascular changes.

ETIOLOGY.—The causes of this form of nephritis are to be sought in some obstruction to the outflow of the renal secretion. It is identical in its simpler form with nephritis produced by ligature of the ureter (STRAUSS and GERMONT), but this simple condition is apt to be complicated by an acute infective process, starting from inflammation or traumatism in the urinary passages, *e.g.*, cystitis, catheterism. The most common causes are enlarged prostate, stricture and cystitis, uterine and ovarian tumours, pregnancy, pelvic inflammation (pyo-salpinx), tubercle and tumours of the bladder, and procidentia uteri; their degree of frequency being in something like the order given. As in other forms of chronic nephritis, cold may easily set up acute intercurrent attacks.

Sex.—My figures give a much larger proportion of males than females, in the ratio of seven to one. The females include the following cases: Procidentia uteri, one; Hysterectomy for uterine fibroma, one; Removal of appendages for double pyo-salpinx, one; Ovariotomy, one. But as the annual number of special diseases of women treated at the General Hospital is very small, while the proportion of cases in which this lesion is found is large, there is reason to think that the disease is more common in females than the above figures would suggest.

Age.—In accordance with the nature of the causes the disease is much more common after middle life.

MORBID ANATOMY.—The kidneys are usually above normal size, rarely small. The capsule is generally thickened and more or less adherent; the surface is granular, red in colour or more often pale. On section, the organ is often tough, soft and flabby. The cortex is sometimes swollen, often reduced in width; the pelvis is always dilated and contains purulent urine, while its mucous membrane is congested and œdematous. Patches of opaque greyish infiltration are visible invading the medulla and cortex, and hæmorrhagic streaks and spots are often present. When there is tubercular disease of the bladder, tubercles may be found in the kidneys. Both organs do not always present identical appearances. One may look fairly normal and the other be in an advanced state of disease, or one may be enlarged, pale and fatty, while the other is smaller, red and granular. Collections of pus may be found in the cellular tissue around the kidneys. In recent cases, especially in connection with pregnancy, the kidneys may appear only swollen and hyperæmic.

The microscopic appearances are as follow:—

Glomeruli.—Some appear normal; others have their nuclei increased, while others again are in a state of hyaline degeneration. Blood may be found extravasated in the intra-capsular space.

Blood vessels.—The larger vessels show well-marked *endarteritis obliterans* with hypertrophy of their middle and outer coats. The capillaries of the cortex are dilated and full of blood; others, especially those in the boundary zone, have their walls thickened.

Convoluted tubules.—These are generally of normal size. The epithelium is in parts normal; in others it shows the appearances of catarrhal nephritis. Some of the tubules are filled with leucocytes, which infiltrate the basement membrane.

Straight tubules.—In the medulla many of these are widely dilated, to a less extent in the cortex, and they often contain colloid material (casts). The epithelial lining is in places being pushed off by invading leucocytes, which may fill up the lumen of the tube.

Connective tissue.—This is swollen and hyaline. Here and there areas may be seen so thickly crowded with invading leucocytes that the normal structure of the kidney is quite hidden, and in places veritable abscesses are visible. This description is based upon cases complicated by more or less acute interstitial nephritis, set up by cystitis; it is probable that where there is simple obstruction there is no interstitial nucleation (STRAUSS and GERMONT).

Angus Macdonald, in his description of the kidney of eclampsia, found only dilated tubules, with altered epithelium and colloid masses in their cavities. This is no doubt the early stage of the uncomplicated process. If not interfered with and given a persisting cause, *e.g.*, enlarged prostate, it ultimately eventuates in red contracting kidneys, such as are described in the following notes from the pathological register:—

CASE 21.—*Enlarged prostate; cystitis; surgical kidneys.*
H. C——, admitted August 3rd; died August 17th. NECROPSY, August 18th, 1886. A spare, emaciated subject.

Heart 9½ oz.; large milk spot on right ventricle; aorta and coronary arteries atheromatous; right ventricle enlarged; slight chronic endocarditis of mitral valve; muscle dark, a little fibrous in places.

Lungs.—Lower lobes congested and œdematous; upper

lobes emphysematous; general bronchitis. The base of the left lung was deeply congested and the pleura was slightly inflamed.

Liver 52 oz.; capsule a little opaque and thickened; parenchyma pale, fatty and mottled with congestion.

Kidneys 8½ oz.; capsules adherent; *red granular kidneys;* pelves dilated with purulent urine; on the left side the cellular tissue round the lower end of the kidney was infiltrated with pus; ureters dilated near their entrance to the bladder.

Bladder.—Great hypertrophy of walls; mucous membrane much inflamed and discoloured, showing small ulcerated elevations to which phosphatic deposit was adherent; prostate much enlarged, fibrous and indurated; prostatic and membranous urethra shewed signs of chronic inflammation.

SYMPTOMS AND COURSE.—These cases generally come under medical or surgical observation on account of the local trouble which causes the kidney mischief, for in this, as in the red contracting kidney of dyscrasial origin, the symptoms are latent, while any special urinary symptoms are masked by the local disorder. The disease usually becomes noticeable when acute nephritis has supervened, and this brings with it corresponding changes in the clinical phenomena.

The urine.—In pronounced chronic cases the urine is increased in amount to 80 or 100 oz.; it is of low density, generally acid unless it has undergone decomposition, usually turbid from some amount of cystitis being present, and containing a moderate amount of albumen, with a little blood. The deposit is muco-purulent, containing hyaline and granular casts, epithelium, pus cells, blood corpuscles, and often triple phosphates and microorganisms.

When acute nephritis has supervened the urine is diminished, the amount of albumen and blood is increased, the density is higher, and epithelial and blood casts are found in addition to the other elements in the deposit. In the former case the *urea* may be normal in amount, but in the latter it is greatly reduced.

The Heart is hypertrophied in a considerable proportion; out of twenty-seven cases it weighed over 10 oz. in twelve, and the left ventricle is described as hypertrophied in one that weighed 10 oz., which may, therefore, be fairly included; thirteen out of twenty-seven are very nearly 50 per cent. This is confirmed by other observers: in enlarged prostate it was observed in four out of ten cases (JEAN); in other cases in five out of twelve (WEIL); while Féré has noticed its frequent occurrence in the bodies of females with procidentia uteri.

Dropsy is generally absent or only slight, but when acute nephritis occurs or heart failure sets in, dropsy supervenes just as it does in lithæmic kidney.

Uræmia.—In pregnancy uræmic convulsions are common, probably because, as already pointed out, there are other co-operating factors present to produce a dyscrasia. Apart from this condition classical uræmia is not common. The usual form is the typhoid type, with dry tongue and feeble pulse. It is a form in which septicæmia plays an important part. Senator has described the type resembling "diabetic coma" as frequently occurring in cystitis (*Vide* Chapter VI., p. 69).

DIAGNOSIS.—These cases can only be diagnosed when their etiology is thoroughly grasped, and the symptoms looked for carefully in cases where the disease is likely to be present.

Evidences of high arterial tension may be found in the character of the pulse, in the accentuation of the aortic second sound and the doubling of the first sound at the heart's apex. Displacement of the latter outwards indicates cardiac hypertrophy. The urine should be collected for twenty-four hours, its quantity measured, its solids and especially its urea quantitatively estimated, and casts looked for.

PROGNOSIS.—Where the disease is due to pressure only, uncomplicated by cystitis, a good result may be hoped for, provided the obstruction is removable and is not of too long duration. This is exactly the condition in most cases due to pregnancy, hence the satisfactory recoveries so often seen after parturition has taken place. In most other cases the cause is of such long-standing that the prognosis is very bad.

ILLUSTRATIVE CASES.

CASE 22.—*Enlarged prostate; cystitis; ascending nephritis.*

A. B., thirty-five, furnace-man, admitted Oct. 23rd, 1888, complaining of swelling of the legs and face, and pain in the back.

His present illness had lasted only a week; he thought he had caught cold, as his work is very hot, and he was in a draught. The next morning his water was scanty and dark red, and has remained diminished in quantity ever since.

Five years ago he was in the hospital with a similar attack, also brought on by cold. He was ill a month. He could recollect no other illness; he had never had gonorrhœa or gout. He had never had any difficulty or pain in making water or had an instrument passed, or was aware of having any bladder disease. General surroundings at home comfortable; had plenty of good food; drank three pints of beer daily, and at one time more than this. Family history good.

Present condition.—Patient is a well-developed, well-nourished man, with a rather sallow complexion. His face looks puffy, but there is no œdema of the eyelids. There is no œdema of the legs or scrotum, but he has been twenty-four hours in bed. T. 97·5; R. 24; P. 60. Tongue clean at tip, furred posteriorly; teeth defective;

gums sore; breath foul; appetite pretty fair; no discomfort, bad taste, or flatulence after food; bowels regular before admission, but not opened since.

Vertical liver dulness $4\frac{1}{4}$ inches, reaching just below the costal border. Spleen not enlarged. No ascites. Chest well formed; respiration vesicular; percussion note normal. Heart's apex in fifth interspace inside vertical nipple line; first sound feeble at apex; second sound loudly accentuated in the aortic area. Pulse 60, full, not easily compressible.

No headache, giddiness, or affection of special sense except slight deafness. Complains of his two forefingers being "dead and white" with the cold in the mornings. No loss of muscular power or common sensibility.

Ophthalmoscopic appearances negative.

Urine, 24 oz., reddish brown, turbid, faintly alkaline; sp. gr. 1015; urea 1·6 per cent.; loaded with albumen; contains blood and peptone, no sugar; white heavy deposit, which under the microscope appears to be composed of pus cells, and flask-shaped or spherical colonies of micro-organisms enclosed in a capsule.

Some urine was drawn off with a catheter under antiseptic precautions, which was sent to Dr. Crooke, who reported as follows: "Urine acid, straw-coloured, tinged with blood; muco-purulent deposit; urea 1·1 per cent.; albumen 1 per cent.; no sugar; blood; pus; no casts; zoogloea masses of bacteria and micrococci, of which several varieties are present."

Some nutrient gelatine was inoculated with the urine and sent to Prof. Crookshank.

It was noticed that although the bladder was full of water when the catheter was introduced, the patient had had no desire to micturate. The lateral lobes of the prostate were moderately enlarged.

Oct. 28th.—Mr. Barling examined the bladder with the cystoscope, and reported as follows: "Direct cystoscope passed without difficulty; bladder previously washed out and about six to seven oz. of fluid injected. Medium rather turbid, but not bloody. Mucous membrane on trigone and immediate neighbourhood elevated, œdematous and fluffy on surface. Neither ureteral orifice could be seen. Still further back the changes were the same, but less marked. No ulceration was seen. The ordinary distribution of vessels on the mucous membrane was quite obscured." The urine remained in the same state.

On Nov. 1st the bladder was ordered to be washed out with boracic lotion daily.

Nov. 6th.—The bladder has been washed out daily since it was ordered. The urine is now pale greenish yellow, turbid, acid, depositing three-quarters of an inch of pus, containing a trace of blood. Under the microscope the flask-shaped and spherical encapsulated masses are much less numerous, but there are plenty of irregular zoogloea masses. Complains of epigastric pain; tongue furred and brown. Has had flatulence. He has been taking saccharin.

As washing out the bladder seemed to be followed by an increase in the amount of pus it was stopped on the 10th, when the amount of pus at once fell to half, but since then it has varied a good deal.

His legs began to get very œdematous, and this remained in spite of his being kept in bed, and of the large amount of water he was passing, which averaged over 80 oz. daily.

The following is the urine report for Nov. 25th: Urine 88 oz., acid, straw-colour, blood-tinged; sp. gr. 1012; deposits muco-pus; urea ·9 per cent.; albumen ·9 per cent.; blood and pus present; slight acetone reaction; one or two flask and hour-glass shaped colonies enclosed within a membranous envelope; also diffuse distribution of bacilli and mycelium-like threads; some hyaline masses to which large fatty cells adhere as if portions of large hyaline or colloid casts (DR. CROOKE).

He left the Hospital *in statu quo.*

Case 23*. *Cystitis ; bronchitis ; ascending nephritis; septicæmia.*

Martha L——, aged forty-three, housewife, admitted May 26th, 1886, with swelling of legs, shortness of breath on exertion, and cough.

She had had pleurisy ten years ago, but no other illness. There was nothing to be made out of the family history. Her surroundings at home were comfortable; she had been temperate, taking only one half-pint of beer daily at her supper. A strongly built, well-nourished woman; her skin had a slightly yellow tinge, but her conjunctivæ were white. There was slight œdema of the legs. T. 98·6; P. 66. Tongue furred; appetite poor; complained of thirst, of flatulence after food, and of vomiting before admission; bowels confined; abdomen distended; vertical liver dulness one and a half inches in the nipple line; ascites. Heart's apex just inside vertical nipple line, in the fifth interspace; there was a diastolic murmur best heard to the left of the apex beat. Pulse full, regular. Urine 10 oz.; sp. gr. 1017; acid; straw yellow colour, smoky; deposit reddish brown; urea 83·6 grains; deposit contained epithelial, hyaline and granular casts, red and white blood corpuscles and renal epithelium.

June 1st.—The cardiac murmur had disappeared, and was never heard again. Urine 30 oz.; Temp. 99·6°; P. 68.

3rd.—Vomiting; urine 44 oz.; Temp. 102·6.

4th.—Vomiting continued, the patient was pale and sweating profusely; her tongue was dry; lungs normal; urine had to be drawn off with catheter, 28 oz.; sp. gr. 1026; acid; claret colour; reddish brown deposit; urea 147·84 grains; albumen quarter column; deposit contained blood corpuscles, pus cells, epithelial, granular

---

* Recorded by Mr. F. H. Noott, Clinical Clerk.

and hyaline casts, and renal and vesical epithelium. Evening temp. 104° F. P. 84.

5th. — Still vomiting; hæmaturia persisting; no pain. Temp. 101·6°; Urine 42 oz.

6th.—Still vomiting; sputa blood-stained; impaired resonance at right apex, with a few rhonchi and râles audible. The patient was becoming distinctly more emaciated and anæmic. Urine 34 oz. Temp. 100°.

8th.—Blood examined with the hæmocytometer appeared normal. Urine 44 oz. Temp. 99·4°; P. 72; R. 48.

9th.—Better. Urine 34 oz. Temp. 98°.

10th.—Not so well. Urine 36 oz. Temp. 97°.

11th.—Better. Urine 40 oz. Temp. 97·2°.

12th.—Passed urine without the catheter. There was pain and swelling over the right parotid, causing inability to open the mouth, but fluids were readily swallowed; there was no redness of the skin; tongue dry; bowels confined. Temp. 98·6°. Urine 60 oz.; alkaline; sp. gr. 1012; slightly blood-tinged; whitish viscid deposit; urea 475·2 grains; a cloud of albumen; deposit contained no casts, but red and white blood corpuscles, pus cells, mucous corpuscles, vesical epithelium, with crystals of triple phosphate and urate of ammonia.

13th.—Swelling great, skin over it hot and red. Urine 58 oz. Temp. 104·4°.

14th.—Swelling less, mouth could be opened more freely; breath foul. Urine 40 oz. Temp. 98·4°.

15th. — Swelling going down; vomiting recurred. Urine 38 oz. Temp. 97·2°.

16th.—Vomiting stopped; catheter required again; swelling still less. Urine 40 oz. Temp. 98°.

17th.—Very slight swelling left; no vomiting; a few râles at left apex; no cardiac murmur; catheter still required. Urine 32 oz.; sp. gr. 1013; alkaline; smoky colour; muco-purulent deposit; urea 239 grains;

albumen one-twelfth column ; deposit contained no casts, but pus cells, blood corpuscles, vesical epithelium and triple phosphates. Temp. 97·6.

18th.—Patient looked very ill, and complained of pain in the bowels, nausea and severe diarrhœa. Pulse 160, very feeble; retention of urine ; pain ceased in the course of the day. Temp. 96·8. Urine 14 oz.

19th.—She had passed a fair night, but the diarrhœa persisted. Temp. 96·8 ; P. 132, very feeble. She vomited, and the ejected matter contained 0·5 per cent. of urea. In the course of the day the diarrhœa stopped, and she complained of no pain. Evening temperature 96·2. Urine 10 oz. She died the following morning at 6.25 a.m.

NECROPSY.—*Lungs, right;* œdema and hypostatic congestion of inferior lobe ; diffuse bronchitis and muco-pus in the bronchi. *Left;* pleuritic adhesions of the base to the thoracic wall ; inferior lobe congested and somewhat collapsed. *Heart* 9 oz. ; no valve lesions ; left ventricular wall hypertrophied. *Liver* 50 oz., misshapen ; right lobe much flattened out, its upper surface grooved by the ribs, and its inferior border reaching nearly to the umbilicus. *Spleen* 5 oz., normal. *Kidneys* 7 oz. ; capsules stripped well ; surface greyish red ; stellate veins prominent. On section, several abscesses were found in the cortex of the right kidney ; the surrounding substance seemed infiltrated and of an opaque grey reddish mottled appearance, as though it were the seat of an acute interstitial nephritis. *Right ureter* dilated ; mucous surface bathed with semi-purulent urine. *Bladder;* walls thickened, much dark spotty ecchymosis of the mucous membrane, which was swollen, injected and infiltrated. *Intestines* collapsed ; wall of small intestine intensely injected ; ileo-cæcal valve very œdematous.

TREATMENT.—The *indicatio causalis* is to remove the obstruction, and when this is recent, as in pregnancy, a cure usually results. The propriety of inducing labour must depend upon the urgency of the symptoms. But in most other cases the propriety of operative proceedings is open to grave question, as they are so uniformly fatal. In determining this difficult question much must depend upon the nature of the obstruction, and also upon the urgency of the local symptoms which the operation is designed to remove. If there are any reasonable grounds for thinking the obstruction has existed not more than a year the chances of a favourable result are improved.

When ascending nephritis is present the use of catheters is attended with great danger, for this is the condition in which the so-called "catheter fever" is likely to be induced.

The general management and treatment of these cases should be conducted on the lines laid down in the previous chapter.

BIBLIOGRAPHY.

FÉRÉ (C.). Note sur les lésions des organes urinaires consécutives à la chute de l'utérus. "Le Progrès Méd." Vol. XII., 1884, p. 22.

JEAN. Thèse de Paris, 1879, p. 59.

MACDONALD (ANGUS.) The Bearings of Chronic Disease of the Heart upon Pregnancy, Parturition and Childbed, with Papers on Puerperal Pleuro-pneumonia and Eclampsia. London, 1878.

STRAUSS and GERMONT. Des lésions histologiques du Rein, chez le cobaye, à la suite de la ligature de l'uretère. "Archives de Phys.," 1882, p. 386.

WEIL. Thèse de Lyon, 1882.

Chapter XIV.

THE COMPLICATIONS OF CHRONIC BRIGHT'S DISEASE.

The complications of chronic Bright's disease are so numerous and important that they require a chapter to themselves.

An analysis of a hundred cases of chronic Bright's disease, taken from the pathological registers of the General Hospital, gives the following as the morbid appearances met with:—

	NATURE OF LESION.	NUMBER.
Brain	Congestion	2
	Meningitis	3
	Œdema	7
	Hæmorrhage	8
Lungs	Pleuritic adhesion	10
	,, effusion	17
	Congestion and œdema	56
	Bronchitis	6
	Emphysema	7
	Infarcts	2
	Phthisis	7
	Pneumonia	19
	Œdema of glottis	1
Heart	Pericardial adhesion	4
	,, effusion	9
	Pericarditis	1
	Fatty heart	7
	Thickening of valves	16
	Hypertrophy	60
	Mitral stenosis	9
	,, incompetence	1
	Aortic ,,	9
	Aneurism of aorta	2
	Atheroma of aorta or coronary arteries	30

Nature of Lesion		Number
Liver	Fatty infiltration	29
	Cirrhosis	13
	Lardaceous degeneration	2
	Nutmeg liver	6
	Atrophy	3
	Cancer	2
	Hydatid	1
	Abscess	1
	Gall-stones	6
Spleen	Capsule thickened	2
	Infarcts	2
Peritonæum	Ascites	2
	Peritonitis	9
	Thickening	1
Stomach	Catarrh	2
Intestine	Tubercular ulcers	1
	Dysenteric „	1
Skin	Cellulitis	4
	Purpura	3
	Eczema	1

URÆMIA.—The most important complications are those classed together under this convenient name. Uræmic vomiting and convulsions are common in the acute febrile nephritis of children, but are not of very serious significance. In the chronic form they are always grave, yet temporary recovery may take place in the most unfavourable circumstances.

Skin eruptions occur sometimes. Rosenstein has described one resembling measles; *erysipelas* and *erythema* are common. *Eczema* sometimes occurs.

Hyperæsthesia of the skin, snatching and burning sensations are occasionally present.

The peculiar condition called "*dead fingers*," is considered by Dieulafoy to be a uræmic symptom, but this is probably an error; it certainly occurs in persons in perfect health.

Deposits of crystals of urea on the skin have been noticed by many trustworthy observers, amongst others by Bartels.

Symmetrical gangrene has been observed in uræmia by Debove.

Vomiting and *diarrhœa* are evidently caused by the vicarious elimination of urinary constituents by the gastro-intestinal mucous membrane, chiefly urea, which is transformed into carbonate of ammonia. It is also probable that the vomiting is often cerebral. In children, vomiting, evidently uræmic and probably cerebral, is very common in acute nephritis.

Hiccough may be very persistent, as in the following case, where it was relieved by the simple method of irritating the anterior nares with a rolled-up piece of blotting paper, suggested by G. A. Gibson.

CASE 24.* *Uræmia; hiccough; peculiar dropsy; peculiar albumen.*

Joseph B., aged thirty-five, slater, was admitted into the General Hospital on Nov. 20th, 1886, complaining of swelling of the face, sickness and headache. About five years ago he first had attacks of sickness and headache, with sharp pains in the left loin and in the groins, not shooting down the leg or into the testicle, but apparently in the course of the colon; he was an in-patient in this hospital at that time, and again in the spring of the present year. On the latter occasion there was no œdema; his chief troubles were headache and vomiting. Since his discharge his face has swollen occasionally; two weeks ago he was very sick and giddy, and in a similar attack he was admitted.

On admission the swelling of his eyelids was so great that he could hardly see out of them. The skin of his face was swollen; no œdema elsewhere. He was a well-developed and fairly well-nourished man.

---
* Recorded by Mr. Pogson, Clinical Clerk.

Pulse, 96, hard, full and regular; R. 16; T. 95·4. Tongue large, pale, and indented at edges. No pain after food; vomiting at times; bowels regular; liver and spleen normal; no ascites.

Heart's apex in fifth intercostal space inside nipple line; a faint systolic murmur with first sound at apex; aortic second sound louder than pulmonary.

Lungs resonant; breath and voice sounds normal.

No pain or difficulty in micturition; urine 30 oz.; acid; sp. gr. 1012; pale straw colour; very slight deposit; urea 1 per cent.; albumen quarter column; no blood; hyaline, granular and epithelial casts, leucocytes and squamous epithelium.

Nov. 22nd.—Eyelids more swollen and red, looking as if he had erysipelas.

Nov. 23rd.—Vomited at 8.30 a.m. Œdema of eyelids *quite pink* from injection, but less in amount.

Nov. 27th.—Œdema of eyelids less; vomiting again.

Nov. 29th.—Not so well; *much troubled with hiccough*.

Dec. 2nd.—Swelling of face less. Frequent vomiting.

Dec. 3rd.—No sleep last night. Frequent vomiting. *Hiccough, removed by irritating nostrils with a rolled-up piece of blotting paper.*

Dec. 6th.—Vomiting still. A uræmic fit at 3 a.m.; another at 3.45, in which he died.

When in hospital previously his urine on one occasion presented the following characters:

April 13th, 1886.—Urine 70 oz.; sp. gr. 1018; alkaline. On boiling and adding acetic acid there was a faint haze. On boiling and adding nitric acid there was no change. On adding a good deal of nitric acid and then boiling there was a haze. On adding nitric acid in the cold there was a dense precipitate. With Esbach's solution the quantity was estimated at 0·1 per cent.

Dyspnœa occurs in uræmia in various forms. Howard enumerates the following:

1. Continuous dyspnœa.
2. Paroxysmal dyspnœa.
3. Both types alternating.
4. Cheyne-Stokes breathing.

But to these we may add a form described by Lecorche as *Laryngeal*, in which the respiration is noisy and sibilant as in croup, attributed by him to spasm of the laryngeal muscles; and the type described in the case of Harriet B., in which the breathing is slow and deep. The most characteristic form is that usually called uræmic asthma, which may be continuous or paroxysmal, as stated by Howard. The patient sits up in bed struggling for breath as in true asthma.

Case 25.—*Chronic lithæmic nephritis; Cheyne-Stokes breathing; recovery.*

George B——, aged fifty-nine, railway servant, was admitted into the General Hospital on June 4th, 1886, complaining of shortness of breathing and swelling of his legs.

His present illness began three months ago with pains in his side, and his feet and legs swelled a bit. This passed off in a week or two, but came on again and lasted till the present time. His previous health had been good; he had never had rheumatism, rheumatic fever or gout. His father and mother died at advanced ages, and all his brothers and sisters are alive and in good health.

He was always fond of beer, drinking four or five pints daily, but he was obliged to keep sober on account of his work. He used to get up twice at night to make water and filled the vessel three parts full, but since his illness this had passed off.

Patient was a well-developed, ill-nourished man; face ruddy; conjunctivæ muddy; some œdema and cyanosis of skin of legs; abdomen distended, measuring thirty-seven and a half inches at umbilicus. P. 90; R. 22; T. 98·2. Tongue clean; appetite good; no discomfort after food; no vomiting; bowels not open, but have been regular; liver enlarged, measuring nine inches of vertical dulness in mammillary line; spleen not enlarged; some ascites. Slight cough, watery mucous expectoration; dulness, deficient vocal fremitus and breath sounds at both bases, more on the right side, with exaggerated breathing above the level of dulness. Heart's apex diffused, but extending to sixth interspace; a loud systolic murmur at apex; pulmonary second sound accentuated.

No difficulty or pain in micturition; quantity 16 oz.; sp. gr. 1020; deep orange colour; slight cloudy deposit; acid; faint trace of albumen; no sugar; no blood; hyaline and fatty epithelial casts, leucocytes and oxalates.

June 5th at 11 p.m. well marked Cheyne-Stokes breathing; average length of pause 20½ seconds; varying from fifteen to twenty-five seconds; out of nine observations, in six it was twenty seconds; active respiration lasted on an average 37½ seconds; in six out of nine observations it was thirty-five seconds, in which period there were about twenty-three respirations. The pulse got slower towards the end of the pause, and beats fuller during descent of respiration.

June 10th.—Had a fairly quiet night, but two bad attacks of dyspnœa after getting out of bed to micturate.

June 11th.—10 p.m. worse; weaker; Cheyne-Stokes breathing less marked.

June 13th.—Abdomen tapped and 64 oz. drawn off.

June 16th.—Cheyne-Stokes breathing again; pause lasted fifteen to twenty seconds; respiratory activity forty to forty-five seconds.

June 20.—Had an attack of dyspnœa last night.

After this he improved and was discharged on July 16th.

Giddiness.—This is a very common symptom, and is

often the first one to be complained of. It is very alarming, and as it is generally persistent it is not easily tolerated, though the patient usually at first attributes it to some temporary "biliousness." It differs, however, from stomach giddiness in being relieved by lying down.

Headache.—The characteristic headache of uræmia is occipital, but it is sometimes frontal. I have not observed any special connection, such as Bartels recognised, between migraine and Bright's disease.

Uræmic deafness has been described by Dieulafoy, and cases have been recorded by Rayer, Rosenstein, Dommergues, Gurowitsch, Lecorché, Eichhorst and Downie. The last observer suggests that it is due to minute hæmorrhages in the cochlea, but it is much more probable that the auditory centre like the visual centre is paralysed by the poison.

Uræmic blindness is a purely cerebral phenomenon without any definite ophthalmoscopic changes. The loss of sight may be complete or incomplete; it is usually transient, lasting from twenty-four to thirty-six hours, but Förster has described a case which lasted seventeen days and terminated in recovery.

Hemiopia sometimes occurs.

Transient hemiplegia without palpable localised lesion is not very uncommon, and cases have been recorded by Paetsch and von Jaeckel.

Delirium is not a very common complication. It is often preceded by headache, visual disturbance and mental confusion, and is generally quiet, but Wagner has recorded a case in which each convulsive attack was followed by violent mania with a temperature of 107·4.

Scholz has described a form of chronic delirium, characterised by hallucinations of sight, delusions of persecution, convulsions and vomiting without headache.

Haslund has published a case of furious delirium with hallucinations which lasted four months. The patient was tormented by hallucinations, and his voracity and salacity were extreme. He died collapsed.

Lecorché and Talamon give a case of a maniac in whom the delirium came on with the disappearance of polyuria and ceased as that returned.

Amnesia with loss of words has been described by Brieger; the case recovered after some time.

CASE 26.—*Chronic lithæmic nephritis (?). Urœmic delirium. Death. Autopsy.*

George R——, aged eighteen, was admitted into the General Hospital on October 6th, 1886, complaining of frontal headache, sickness, loss of appetite, and making a lot of water at night.

He said he did not feel weak and did not think he had lost flesh. Four months ago his illness began with eight fits, extending over two nights and a day. He had had no fits before or since. He had never had any illness or accident, except having his head cut open by a swing-boat eight years ago. He had never had dropsy, any swelling, scarlatina or gonorrhœa.

He was a badly nourished, ill-developed, sallow-faced lad. P. 100, small, weak, regular; T. 98·8; R. 24. No œdema.

Chest symmetrically formed; expansion equal; percussion note resonant; breath sounds, vocal fremitus and resonance normal.

Heart's apex in fifth interspace inside the nipple line; impulse variable; action regular; no thrill; sounds at apex normal; in tricuspid and pulmonary areas there was a systolic blowing murmur; the aortic second sound was accentuated; there was no murmur in the neck.

Tongue flabby, indented at edges, pale, but clean; bowels regular; no discomfort after food; appetite poor. Liver and spleen normal; no pain or difficulty in making water, but gets out of bed every half hour to make it; quantity 54 oz.; sp. gr. 1012; alkaline; greenish yellow; deposits white flocculent precipitate; urea ·8 per cent.; albumen half column; no sugar or blood. Under the microscope groups of leucocytes, hyaline and granular casts, red blood corpuscles, squamous and renal epithelium, triple phosphates, granular *débris* and masses of micrococci were detected.

He complained of dimness of vision, which had come on for twelve or thirteen weeks, being only able to read large capitals. With the ophthalmoscope extensive neuro-retinitis was seen in both eyes, with large white glistening patches of retinal atrophy; both eyes were equally bad.

Oct. 6th.—Complained of fixed pain in the left side; bowels confined; tongue foul.

7th.—Vomited about two pints of brownish watery matter, free from blood; bowels acted loosely; tongue very brown. Complained of sick headache all last evening.

9th.—In the afternoon was restless, thirsty and faint. He vomited a good deal, and was given some brandy, after which he improved.

10th.—Had a sleepless night, but was quiet; complained of headache and chilliness; vomited five times during the night, 15 oz. in all, which contained 0·15 per cent. of urea.

11th.—*Has been whistling and singing most of the night;* slept only half an hour; kicked the clothes off continually; complains still of headache. Vomiting again this morning; vomit gives the peptone reaction with Fehling's solution, and turns methyl-violet slightly blue.

12th.—Slept only from 12 to 12.30 a.m.; *restless, noisy and delirious;* he continued delirious all day, and towards evening got out of bed; he slept with a chloral and bromide draught. To-day, some dulness, friction, deficient breathing and coarse crepitation were noticed over right side of chest anteriorly. His temperature continued to be normal or subnormal. Pulse 120; Respirations 36 to 48. He died the following day at 1.15 p.m. without any development of fresh symptoms.

Autopsy, October 14th.—*External appearances.*—Features pale; lips bluish; *rigor mortis* present; hypostasis more marked posteriorly.

Head.—No trace of previous injury to skull; brain congested; excess of serous fluid in sub-arachnoid space and ventricles; no gross lesion in any part of brain substance.

Heart.— Pericardial cavity contained 4 oz. serous fluid; heart 13 oz.; generally hypertrophied, especially left ventricle; muscular wall firm and rigid; valves normal.

Lungs both adherent by old fibrous adhesions; old caseous and fibroid phthisis at left apex. In centre of right middle lobe was a small hard nodule, composed of caseous material, surrounded by a capsule of pigmented fibrous tissue, while a similar, but rather larger nodule was in the upper lobe of the same lung. The bronchial glands were enlarged, caseous and encapsuled in fibrous tissue; the left lung contained several smaller caseous nodules of the same kind.

Liver 51 oz.; pale, soft and fatty. *Spleen* 5 oz.

Kidneys, right, 5½ oz., enlarged, surface greyish red mottled with yellow spots; capsule slightly adherent; cortex increased, swollen, greyish with yellow mottling and hyperæmia intermingled, translucent-looking in places and generally fatty; medullary cones congested, almost claret-coloured; *left*, 4 oz., smaller, more granular, capsule more adherent.

Among the minor phenomena of threatening uræmia, *Cramps* are common. Charcot has seen a kind of *tremor* resembling paralysis agitans, and Jaccoud has described *tonic spasm* of the flexors of the forearm and the posterior muscles of the neck.

Headache is a very common prodromal symptom; it may be frontal or occipital, but the latter is more characteristic.

Twitchings are also a common early symptom, and these are closely related to *convulsions*, which may attack single groups of muscles, and to Jacksonian epilepsy, or convulsions without loss of consciousness, which may be unilateral (TENNESON and CHANTEMESSE).

Convulsions of the epileptic type constitute the common uræmic fit, being accompanied by total loss of consciousness, biting the tongue, and foaming at the mouth, generally followed by deep coma. The attack may consist of a single paroxysm or a succession of these, the patient lying unconscious in the intervals, breathing stertorously, or in a sort of deep sleep from which he may be partially roused.

Coma may be the result of a convulsive attack, or may come on gradually, or suddenly as in apoplexy. Roberts has given several cases of this sudden onset.

The *temperature* during the attack is generally subnormal. Roberts has recorded a case in which the temperature fell as low as $94.4°$ F., and Hirtz another in which it fell to $93.9°$, while Bourneville has published temperatures as low as $93.1°$, $91.1°$, $89.6°$, and even $86.1°$. According to Hutinel these low temperatures are more often met with in those forms of nephritis following diseases of the urinary passages, *e.g.*, surgical kidney, and Netter observed a temperature of $86°$ in a case of anuria due to purulent nephritis with a twist of the left ureter.

MacBride gives the following conditions as those under which lowering of the temperature is always present: (1,) In kidney diseases following affections of the urinary passages with complete obstruction; (2,) In uræmia in old persons; (3,) In uræmia occurring in old standing kidney diseases complicated by vomiting, diarrhœa and hæmorrhages; (4,) In uræmia which supervenes in cancerous cachexia and marasmus. In other cases elevation of temperature has been observed even as high as 105·8° and 107·4°, but this has always been followed by a rapid fall.

The *pulse* is generally quick, but may be as low as 60 or even 40.

Case 27.—*Chronic lithæmic nephritis; uræmic coma. Recovery.*

Richard J——, aged forty, brewer, was admitted into the General Hospital on March 20th, 1886, complaining of swelling, which began in his legs not in his face, and at the same time he noticed that his urine was smoky and diminished in quantity. His previous health had been good; he had never had scarlatina or gout; he was ridden over about two years ago and got "concussion of the spine," for which he was an in-patient in the surgical wards of this hospital. He was not told that his kidneys were hurt.

He had been in the habit of getting out of bed to make water five or six times for the last ten years; he drank about three pints of beer daily, but no spirits as a rule.

He was a stout, well-developed man, with general anasarca, but no fluid in pleuræ or peritonæum. Pulse 84, regular, full; R. 24; T. 98 F.

Tongue clean; fulness after food; bowels loose from purgatives; liver normal; spleen a little enlarged.

Heart's apex in fifth interspace in nipple line; systolic murmur at apex, not constant; aortic second sound accentuated.

Lungs resonant; breath and voice sounds normal.

Urine 44 oz., acid; sp. gr. 1015; smoky, yellow; white deposit; urea 1 per cent.; albumen five-sixths column; blood present; peptones present; fatty, hyaline and epithelial casts; white and red blood corpuscles and renal epithelium.

He complained of dimness of sight; there were subconjunctival hæmorrhages in the right eye. With the ophthalmoscope numerous atrophic patches were seen in both retinæ, and a recent hæmorrhage in the right eye: the outlines of both discs were gone, the arteries were very small, and the veins large.

In spite of purging and diuretics his dropsy increased, and his urea excretion was very little. On March 26th it was 174 grains per diem. He began to complain much of occipital headache and pain behind his eyes.

April 4th.—His legs were drained.

5th.—He began to be drowsy, and passed a stool under him. He continued in this apathetic condition, passing urine and fæces in bed till his tongue grew dry, sordes appeared on his teeth; he was *delirious* and did not seem to recognise people. This continued till April 25th, when he gradually sank into coma. When seen his pupils did not react to light; he could be partially roused; he had vomited several times. The coma seemed to deepen, and an ineffectual attempt was made to bleed him. He was ordered a hot-air bath, in which he sweated freely, and next morning was better. After this he slowly improved, and after being for a time at the Jaffray Hospital, was discharged, free from œdema and fairly well, but quite blind.

HÆMORRHAGES of all kinds are very common. Profuse hæmaturia has been already alluded to, but hæmorrhage may take place from the bronchial or intestinal mucous surfaces, into the retina, the tympanum and the skin, while probably it constitutes the most common cause of cerebral apoplexy. The cause of the hæmorrhage is in all cases the same,—degeneration of the small vessels and increased blood pressure.

CASE 28.—*Chronic lithæmic nephritis; pneumonia; cerebral hæmorrhage. Death. Autopsy.*

Edward C——, aged thirty-two, foreman at bicycle

works; admitted April 9th, 1886, with headache, shivering and pains in the back, worse on the right side and increased on deep inspiration. He had been ill one day.

Except vague complaints of rheumatism, without joint affection, there was no previous illness. His habits were temperate and his home comfortable, but his work exposed him to changes of temperature. He was a well-built and well-nourished man, but looked ill; his face was dusky, and he complained of feeling cold. T. 104; R. 38; P. 120.

There was no cough, and no abnormal physical signs could be detected. There was no œdema. Tongue dry in centre.

Urine 24 oz.; sp. gr. 1013; acid; copious brown deposit; albumen ·1 per cent.; epithelial, hyaline, and blood casts, with blood corpuscles, and renal epithelium in the deposit.

The next day he began to vomit, and his bowels became loose, but the stools were dark. His evening temperature was 103° F. On the 13th he was delirious, and enteric fever was suspected, but there was no rash, and the stools were not at all characteristic. On the 16th dulness and coarse crepitation were made out at the right base posteriorly, and on the 17th his temperature came down to normal, and remained down afterwards. He was some time getting better. The urine continued to be bloody and albuminous, but sufficient in quantity. On May 22nd the urine report is as follows: 56 oz.; sp. gr. 1010; acid; pale straw colour; urea ·95 per cent.; albumen ·5 per cent.; a trace of blood; epithelial and hyaline casts, with blood corpuscles, and renal epithelium in the flocculent deposit.

On June 8th he was made an out-patient. He was re-admitted on September 29th with effusion in the pleural cavity. On October 6th œdema was for the first time noted in the legs; this increased, and never left him. His urine contained albumen, and numerous epithelial and hyaline casts, blood and epithelium. He was very noisy and restless for some days, but by October 11th he was quieter. On the 17th pericardial friction was heard. He began to vomit and complained of headache on the 21st. Ascites was noted on the 27th. He remained more than two months in the Hospital, but insisted on going out on December 9th. His legs were still a little œdematous and there was fluid in the peritonæum; the urine was fairly abundant, 52 oz.; slightly albuminous, but in the last examination no casts were seen; urea 1·05 per cent.

He was again admitted on January 8th with great swelling of the abdomen, legs and scrotum. Urine 68 oz.; sp. gr. 1012; pale brown colour; slight white deposit; a quarter column of albumen; a little blood, and no casts. In the morning of January 18th he vomited again, and at 7.40 p.m. when seen he appeared to be in a fit affecting the left side; the left eyelid drooped, and the left arm was rigid, with occasional spasmodic jerkings; he could not speak, but made intelligible signs with his right hand. His teeth were clenched, and now and then ground firmly together. His breathing was noisy and occasionally stertorous. Vomiting came on, and he died at 9.20 the same evening.

AUTOPSY, January 19th, 1887.—Body of a fairly nourished man, *rigor mortis* and *post mortem* staining well marked. Legs and feet œdematous, abdomen greatly distended with a large quantity of clear straw-coloured serum. Pericardium containing from 1 to 2 oz. of blood-stained fluid; there was also a small quantity of lymph on the surface of the heart and serous surface of the pericardium.

Heart 1 lb., hypertrophied, wall of left ventricle ¾ in. thick, muscle substance hard, firm and pale. Cavities not dilated. Musculi papillares and columnæ carneæ well developed; aortic and pulmonary valves normal; mitral orifice admitted three fingers; tricuspid, only the tips of five fingers; the posterior cusp mitral valve shortened but not thickened.

Lungs.—*Left* pleura contained a large quantity of clear straw-coloured fluid; the parietal pleura was much thickened, and the lung bound down to the spine by firm adhesions, the lower lobe was entirely collapsed and attached to the diaphragm by a short tough fibrinous band about the thickness of the little finger. The upper lobe was crepitant throughout, but much firmer than normal. The whole lung weighed 8½ oz., and had somewhat the appearance of a Florence flask, the lower lobe forming the neck.

Right lung 1 lb. 4 oz.; this was adherent, specially the lower lobe, the surface of the lung being torn in removing it; the tissue was firm and congested.

Liver 3 lbs. 2 oz. *Spleen* 4½ oz. *Kidneys*, right, 4 oz.; left, 3½ oz., they were small and pale, and on section were found to be hard and fibrous, cutting grittily, the cortex narrowed, of a pale yellowish white colour, hard and fibrous; the pyramids were pale, also hard and fibrous, the capsule thickened and adherent, on removing it the surface was mammillated.

Bladder was empty and contracted, nothing abnormal was noticed.

On removing the *brain* the convolutions on the (left?) side were flattened, the (left?) lateral ventricle was much distended with blood and clots, and some blood had found its way into the ventricle on the other side, the corpus striatum and optic thalamus were much lacerated and disorganised by the hæmorrhage.

Case 29.*—*Chronic lithæmic nephritis; pemphigus; purpura hæmorrhagica. Death. Autopsy.*

---

* Case recorded by Dr. Stacey Wilson, House Physician.

Ann D——, aged sixty-four, admitted December 13th, 1886, with an eruption all over her body. She said that three weeks previously brown patches appeared on her face and arms, which all at once became scaly and peeled off "like shavings," the scales being renewed. The rash spread over her body, and in the second week blisters formed, chiefly on her hands, which broke and became red and sore; her nose bled, and blood came from her mouth, the gums were very sore; her water was bloody, and her motions dark, and there was a discharge from both her ears.

She had had no previous eruption; when a child she had acute rheumatism. No history of syphilis.

She could give no details of her family history, except that her father and mother died young, she thought of "fever," and that her only sister was alive and well.

On admission her face was covered with crusts and dried blood. There was blood on her gums and tongue, and under the latter were small ulcers. Her whole body was covered with a hæmorrhagic eruption of spots as large as a pin's head; in places these had coalesced to form larger patches. There were numerous blisters the size of three-penny pieces on the arms, which were covered with scaly scabs. In the flexures of the elbows, knees, groins, and axillæ, as well as on the shoulders and buttocks, the skin had lost its epidermis and bled freely. A very offensive smell arose from the patient. The following day the whole of the epidermis from the sides and soles of the feet came away, leaving raw bleeding surfaces. On the 16th, the third day after admission, she rapidly sank and died.

AUTOPSY, December 17th.—*External appearances.*— A well nourished woman, with well marked *post mortem* rigidity and hypostatic congestion. The entire body was to a greater or less extent marked with either small round deeply stained hæmorrhagic spots, or large irregular red patches of denuded epidermis, or actual ulceration. On the face was a mixture of spots, of varying size and depth of colour, and small red patches. The mucous membrane of the gums was rough and irregular, but no hæmorrhages could be seen. The spots were most abundant on the shins, calves and abdomen; the patches most on the buttocks, backs of hands and feet, flexures of elbows and knees, groins and vulva.

Brain.—Basilar artery atheromatous; dilatation of vessels of pons and medulla; otherwise normal.

Lungs.—Slightly adherent; some hæmorrhages on the diaphragmatic pleura, the largest the size of a sixpenny piece.

Heart.—Several small hæmorrhages in cellular tissue at roots of great vessels outside epicardium; weight 11 oz.; much fat on surface; muscular wall soft and thin. Valves healthy, except a little atheroma on the mitral valve. No sub-endocardial hæmorrhage. Aorta dilated and slightly atheromatous.

Liver 46 oz.; soft, congested and fatty.

Gall-bladder dilated, adherent to transverse colon; contained two white mulberry calculi, and 1½ oz. of dark thick bile.

Kidneys 9 oz.; capsules adherent; surfaces granular; cortices narrow and hard.

Spleen 2 oz.; normal.

Intestines.—There was a hæmorrhage the size of a shilling in the wall of the ascending colon, about three inches above the cæcum, and in the mesentery there was a small cyst containing ½ oz. of clear fluid. The *psoas and iliacus* muscles on both sides contained hæmorrhages, more on the right side. No other hæmorrhages were observed in any situation.

CHRONIC PACHYMENINGITIS.—I take the following from the report of the *post mortem* examination of a man named J. C——, in the Pathological Register of the General Hospital.

Brain.—*Dura mater* adherent to the calvarium and in part to the surface of the brain; the meninges were generally thickened all over the vertex, œdematous, and the *pia mater* was here and there adherent to the vertex, the grey matter of which appeared dull, cloudy and mottled with yellow. There was much serous effusion in the sub-arachnoid space and ventricles. A small "psammoma" was found adherent to the membranes behind the left crus. The under surface of the left temporo-sphenoidal lobe was broken down with red and yellow softening; the corpora striata and optic thalami were cloudy and mottled yellow and pinkish grey. In the softened area the small vessels shewed with the microscope fatty degeneration of their walls. *Heart* 13 oz.; left ventricle considerably hypertrophied; no valve lesion. *Kidneys* 9 oz.; atrophied, capsules adherent; surfaces granular; cortices visibly diminished and fibroid.

ŒDEMA OF THE GLOTTIS.—This is a well known complication of general dropsy in acute and sub-acute nephritis, but it is not commonly known to occur in chronic Bright's disease.

A man was brought into the hospital dead or dying, and on *post mortem* examination there was well marked œdema of the glottis, without general œdema. He had typical contracting kidneys; his heart weighed $13\frac{1}{2}$ oz., its valves were thickened, and the coronary arteries atheromatous; his lungs were congested and œdematous; his liver was cirrhotic and his brain œdematous.

The *larynx* is not uncommonly the seat of acute or chronic catarrh.

CONGESTION AND ŒDEMA OF LUNGS.—Among the most common *post mortem* appearances in persons dying with contracting kidney, whatever the cause of death may have been, are congestion and œdema of the lungs.

This is the condition which during life probably gives rise to the symptoms of bronchitis which are so common in these patients. Many elderly patients who complain of "bronchitis," and who have cough and difficulty of breathing, will be found on examination to have no physical signs sufficient to account for their complaint. If the case is gone into and the urine examined sufficient evidence will often be found to enable it to be recognised as a latent example of contracting kidney.

One of the fatal complications which may arise in the course of latent contracting kidneys is *acute œdema of the lungs*. Its explanation is not at all clear, but it is generally attributed in a loose sort of way to the dyscrasia. This is all very well if there is dropsy elsewhere, but is no sort of explanation for the sudden development of dropsy in the lungs when all the rest of the body is free. Acute œdema of the lungs is too rare a condition to have attracted much attention. Standard text books either leave it out altogether or refer to the unimportant forms accompanying pneumonia and bronchitis, or to the pulmonary œdema of mitral disease.

In seeking an explanation for it we must apply the principles laid down in the chapter on Dropsy. Gradual heart failure would not give rise to these sudden attacks of œdema. They are not like the creeping dropsy of failing heart, but resemble the outburst of anasarca in acute nephritis. We must look therefore to some causes favouring the rapid outpouring of lymph. These may perhaps be found in hypertrophy of the right side of the heart and in increased permeability of the capillaries due to some local poison.

In the case about to be related œdema of the lungs followed an anginal attack, in which the pulse was not affected. Could the angina have been due to some affection of the right side of the heart? Could there be spasm of the pulmonary vessels giving rise to heart pain, and followed by paralytic distension and œdema? These are possibilities.

Case 30.—*Chronic lithæmic nephritis; uræmia; death from œdema of lung.*

A lady, aged sixty, of gouty antecedents, eight years before had had an epileptiform seizure, recognised at the time to be " uræmic." No other kidney symptoms had developed, except the necessity to rise at night to pass water. For the last few weeks there had been a little dry cough, with some dyspnœa on exertion, and a complaint of pain in the chest.

One day, after a short walk, she was seized with a violent attack of pain at mid-sternum, which went through to her back and down the left arm to the *thumb*.

When seen by me at 2.30 p.m. her face was pale and anxious, extremities cold. The heart's apex was in the sixth left interspace, an inch outside the vertical mammillary line; the sounds were very feeble. The pulse was full, soft and regular, but she had been inhaling nitrite of amyl. Respiration was easy; slight cough; no expectoration; breath sounds normal. She complained a good deal of the pain, which was not relieved by any remedy. By 4.30, when I left her, her face was less anxious, her hands were warmer, and her pulse was still full and quite soft. She continued better and sat up a little during the afternoon, but about eight o'clock difficulty of breathing commenced. When I returned at 9.30 there was marked œdema of both lungs, moist bubbling râles all over the chest, cyanosis of the face, cold clammy skin, &c.

All remedies were useless; she gradually became unconscious, and died at 11.30. No urine was passed after the first onset in the middle of the day, when, at the same time the bowels were moved. The pulse remained full till quite near the end.

Bronchitis accompanied by emphysema is found *post*

mortem with some frequency, but, as already hinted, not so frequently as the cough and dyspnœa from which these patients suffer would suggest.

CASE 31.—*Chronic lithæmic nephritis; gout; bronchitis; eczema; uræmic delirium. Death. Autopsy.*

James W——, aged sixty-five, upholsterer, admitted January 10th, 1888, with cough, dyspnœa on exertion, and a breaking out on his skin.

He had had a cough all the winter, but the eruption came out eight weeks ago. Twenty years before he had a slight attack of acute rheumatism, he was ill only a week; five years ago he had an attack of gout, and for some winters past he had suffered from bronchitis. He had lately been badly off for food, but he had taken one to two pints of beer daily.

He was a well developed, poorly nourished man; the skin of his head, face, and forearms was intensely red, and covered with crusts; he complained a great deal of burning and itching. T. 100; P. 96; R. 24. Tongue furred; bowels confined. Heart sounds feeble, but no murmur; pulse weak and compressible. Breath sounds feeble anteriorly; posteriorly, harsh and accompanied by moist râles.

Urine 50 oz.; sp. gr. 1026; contained a trace of albumen; uratic deposit.

After admission he made no improvement.

January 23rd.—T. 100; P. 96; R. 42. He had been delirious, getting out of bed and asking where he was; urine 50 oz.; sp. gr. 1022; urea 407 grains; a thick cloud of albumen; a trace of blood; hyaline and granular casts.

27th.— T. 102·8; P. 102; R. 36. Still delirious and getting out of bed.

28th.—T. 101; P. 108; R. 50. Quieter; towards evening increased difficulty of breathing.

29th.—Had passed a bad night with his breathing;

his respirations were sighing and laboured; expiration was greatly prolonged. T. 102; P. 90; R.34.

31st.—Pain, redness and swelling over right hip joint. T. 100·5; P. 100; R. 48.

February 1st.—Breathing noisy from rhonchi; hip very red and swollen. T. 102·5; P. 98; R. 46.

2nd.—Towards evening breathing very short. T. 102; P. 100; R. 52. Urine 28 oz.; sp. gr. 1020; acid, straw colour, smoky; white flocculent precipitate; very thick cloud of albumen; deposit contained cellular hyaline and epithelial casts, blood corpuscles and blood casts, and renal epithelium.

3rd.—Breath very short; hip much inflamed. Death at 2.50 p.m.

Autopsy, February 4th, 1888.—*Abstract of notes:*—There was a large abscess over the hip joint, connected with dead bone. The lungs shewed chronic bronchitis and emphysema, with dilated bronchi. The pericardium was adherent, and the heart hypertrophied, weight 15 oz. The kidneys were typical red granular atrophic kidneys, weight 8 oz. The arteries in the boundary zone were thickened and patent. There were numerous yellow spots in the cortex, which were composed of urate of soda. In the metatarso-phalangeal joints of both great toes there was chronic arthritis, with deposit of urate of soda in the fibrous tissues.

Pneumonia.—It is very common to find contracting kidneys in the bodies of persons dying of acute lobar pneumonia; eighteen out of a hundred cases of chronic Bright's disease collected for statistical purposes had this association. No doubt the presence of such a condition is a most unfavourable element in the prognosis of acute pneumonia. Lobular pneumonia may occur in connection with bronchitis, and hypostatic pneumonia may supervene in the later stages of slowly dying cases.

CASE. 32*. *Bronchitis; chronic lithæmic nephritis; lobular pneumonia. Death. Autopsy.*

J. J——, male, sixty-seven, chaffcutter, admitted September 20th, 1886, with cough and shortness of breath.

Family history unimportant. Had rheumatic fever twenty-four years ago, and has since suffered frequently from rheumatism, colds and cough. Admits that he has not been always temperate, and has frequently been insufficiently fed. Three weeks ago he caught cold, but getting worse, and a sharp pain attacking his left side, he came to the Hospital.

A decrepit old man, lame from rheumatic stiffness of the hip; face pale; conjunctivæ jaundiced; respiration laboured. T. 100. Lips blue; no teeth; tongue pale, moist, furred at sides; appetite bad; constant thirst; throat dry and parched. Has pain after food; eructations of wind; water brash; bowels confined. Liver projects an inch below ribs. Inguinal hernia on right side. Heart sounds reduplicated in pulmonary and tricuspid areas; apex beat not to be felt, but dulness does not extend outside nipple line. Sputa frothy; cough troublesome; respirations 30; breathing harsh, expiration prolonged, accompanied by musical rhonchi all over chest. Gets up several times at night to make water.

Urine 20 oz.; acid; sp. gr. 1015; deep amber, smoky; white flocculent precipitate; faint cloud of albumen; blood present; no sugar; under microscope hyaline casts, a few red blood corpuscles and leucocytes, free renal epithelium, spheroidal and polymorphous epithelium from urinary tract, squamous epithelium, spermatozoa, and uric acid.

Progress of Case.—The cyanosis increased; he was troubled at night with asthma-like attacks of dyspnœa.

---

* Recorded by Dr. Lewis Hawkes, Acting House Physician.

By the end of September he began to be drowsy. His temperature varied, but never exceeded 101°. His bowels were opened regularly, and were rather loose; The urine was usually scanty, about 20 oz. in twenty-four hours. The drowsiness increased, and he died comatose on October 6th.

AUTOPSY, October 7th.—*Heart* 14 oz.; right side full of clots; left ventricle empty, contracted; valves thickened, but competent; left ventricle hypertrophied; heart's muscle generally pale, soft, and friable.

Lungs.—Turbid yellow serum in pleural cavities; right, 300 ccs., left, 100 ccs. Right lung adherent to diaphragm. Both lungs emphysematous in front, œdematous, congested, and in parts collapsed posteriorly. *Scattered through the left lung* were patches as large as a bean, some fresh and vascular, others pale greyish and gelatinous-looking, trabeculated and puriform in the centre, which under the microscope shewed pulmonary alveoli filled with fibrin, round cells and desquamated epithelium. In both lungs there was generalised bronchitis, with dilatation of the larger bronchi and peribronchial thickening.

Liver 42 oz.; capsule thickened and adherent to diaphragm. On section, congested, soft and friable.

Spleen 3½ oz.; capsule thickened.

Kidneys 8½ oz.; both kidneys atrophied; capsules adherent; surfaces granular. On section, cortex diminished, of a brown red colour, dotted with grey and yellow points; medullary portions of a darker hue.

THE ALIMENTARY SYSTEM.—*Catarrh of the stomach* is a very common complication. It is generally not very intense, giving rise to flatulence, weight after food, and attended by morning sickness or nausea. There is often some chronic pharyngeal catarrh associated with it. The bowels are generally *confined*, and the action of the liver sluggish. Sometimes constipation alternates with attacks of *diarrhœa* due to intestinal catarrh. *Hæmatemesis* and *hæmorrhage* from the bowel occur occasionally. *Hæmorrhoids* are not uncommon. The *liver* is generally diseased, most commonly fatty, often more or less cirrhosed or congested, or shews signs of lardaceous degeneration; in some cases the liver has undergone simple atrophy.

BIBLIOGRAPHY.

BARTELS (C.). *Op. cit.*, p. 490.
BRIEGER. "Charité-annalen." Bd. VII.
CHARCOT (J. M.). *Op. cit.*, p. 316.
DEBOVE. "Bulletin de la Soc. Méd. des Hôpitaux de Paris." Feb. 27, 1880.
DIEULAFOY. "Union Médicale," Aug. 6, 1882; and "France Médicale," 1877, No. 16.
DOWNIE (J. W.). Deafness in Bright's disease. "Glas. Med. Jour.," Vol. XXIV., 1885, p. 410.
EICHHORST. Handbuch der speciellen Pathologie und Therapie. 1884, Bd. II.
GIBSON (G. A.). A Classical Remedy. "Edin. Med. Jour.," Vol. XXXI., 1886, p. 912.
HASLUND. Quoted by Berger,—Troubles intellectuels dans le cours de le Néphrite chronique. "Arch. Gén. de Méd.," 1880, II., p. 738.
HUTINEL. Les Températures basses centrales. Thèse de Paris, 1880.
JACCOUD. "Clinique de la Charité," 1874, p. 499.
LECORCHÉ and TALAMON. Études médicales, p. 172.
ROSENSTEIN. *Op. cit.* 3rd edition, p. 241, *et seq.*
TENNESON and CHANTEMESSE. "Bull. de la Soc. des Hôpitaux de Paris," 1869, p. 50.
VON JAECKEL. Beiträge zum Symptomencomplex der Urämie. "Inaug. Diss.," Berlin, 1884.
WEST (S.). On the occurrence of Blood in the Urine in granular Kidney. "Lancet," 1885, II., p. 104.

Chapter XV.
THE TREATMENT OF LITHÆMIC NEPHRITIS.

Successful treatment depends in the first place upon a correct understanding of the principles of diagnosis and prognosis which have been laid down. We are all liable to try to do too much in the way of treatment. This arises from our ignorance of the exact nature and probable course of diseases, and our consequent over-anxiety as to their result.

Fortunately this obstacle is year by year diminishing, as we acquire better knowledge.

If a patient suffering from chronic Bright's disease is in fair general health, and the discovery of the lesion has been made more or less by accident, so to speak, we should be careful to do nothing that will worsen his condition by over vigorous treatment or too severe regimen, but should be content to relieve the symptoms of which he complains, if it is in our power to do so, while at the same time we endeavour to regulate his habits and mode of life in conformity with principles deduced from our knowledge of the etiology and pathology of his disease.

CLIMATE AND HYGIENE.—It is seldom that patients have it in their power to choose their residence in strict conformity with medical advice, but where it is possible they should seek a dry warm climate for the winter. Algiers is uncertain, as the weather is liable to sudden changes from an English summer climate to deluges of rain with cold wind. Egypt is often bitterly cold, but is dry. The uplands of South Africa are a long way off, but as our winter is their summer they afford the invalid who is in fair health and can make the change, the opportunity of enjoying two summers, while residence there all the year round fulfils the required climatic conditions.

During the summer months Europe is less difficult, and a visit to the Engadine or to the higher parts of the Scottish Highlands, for example, to Braemar in Aberdeenshire, or to Buxton in Derbyshire, is greatly calculated to improve the general health, and should be undertaken if possible even by those unable to get away in the winter.

If the patient stay at home, his house should be on a dry soil, and sheltered from the north and northeast winds. Our south coast health resorts, St. Leonards, Ventnor, Bournemouth and Torquay, do not fulfil all the requirements of climate, but offer during the winter, shelter, a comparatively mild climate and a moderate rainfall.

Clothing.—The patient should wear woollen clothing by day and by night, summer and winter, and in all climates. The adoption of the so-called "Jaeger system" in its entirety is perhaps the very best thing these patients can do.

The use of the *hot air* or *vapour bath* in their own rooms once a week is better than taking a Turkish bath away from home. After the bath the patient should be well rubbed down and then go to bed. The duration of the bath should not exceed twenty minutes. The skin may not act well the first few times, but will gradually acquire activity. If the duration of the bath is prolonged it may cause faintness.

Cold bathing or *sponging* should be avoided, but tepid sponging every morning may be allowed.

Daily *exercise* should be encouraged, but violent exercise such as running or anything approaching fatigue, is to be avoided, having in view the state of the heart and vascular system.

DIET.—Abstinence from butcher's meat, cheese, and all alcoholic drinks should be enjoined.

A liberal dietary may be obtained from light meats,

such as sweet-breads, tripe, cowheel, calves' head and feet, fish, fowl, game, eggs, butter, milk, cream, fruit, vegetables and farinaceous foods.

Milk should be taken in moderation, and should not be used as a drink with meals.

Tea, coffee, cocoa and chocolate are permissible.

Sugar and sweet things should be forbidden.

I am convinced that absolute milk diet is in many cases not only unnecessary and extremely distasteful to most patients, but positively harmful, as it does not supply enough nourishment. In some persons it causes obstinate constipation; in others I have seen the urine which was intensely acid during its use improve greatly on a mixed diet being substituted. It must be remembered that the alkalies needed to keep up the alkalinity of the blood are derived from farinaceous food, while milk contains many sources of acidity. When milk is desirable it should be always given with bread or farinaceous food or puddings.

Ordinary drinking water generally contains a considerable quantity of sulphate of lime, at any rate in this district, and I am convinced is very hurtful. One of the best substitutes is Salutaris water; or if there is much acidity, Vichy Water (Célestins or Haute Rive), which may be flavoured with lemon juice, may be substituted. If the ordinary water supply is very hard, distilled water should be used for making tea, coffee, &c.

Beef tea and soups of all kinds should be interdicted; the former has a chemical composition closely resembling that of urine (MASTERMAN), and all soups partake more or less of this character, unless they are made without any stock, as some German soups are, but these are hardly understood by the British cook.

By every means endeavour to keep up the patient's general health and nutrition, and set aside any of the above injunctions which interfere with these all-important considerations.

There may frequently be a difficulty about giving up alcohol, nevertheless its disuse should always have a fair trial; should a certain allowance be absolutely necessary it is best given in the shape of good whiskey well diluted with Vichy, Kronenquelle, or Salutaris water.

As a general prescription, to fulfil the main indications of those cases where no special symptoms are calling for treatment or contra-indicate its use, a mixture containing *sodium benzoate* and *digitalis* may be employed.

  ℞ Sodii Benzoatis   gr. x
   Tincturæ Digitalis   ♏x
   Inf. Gentianæ   ad.   ℨj
   M.
  Sig. To be taken three times a day.

The bowels should be kept acting by some laxative, if they are at all sluggish.

   Euonymini   gr. j
   Ext. Aloes   gr. ij
   Ext. Belladonnæ   gr. ½
   Ft. pil.
  Sig. To be taken at bed time, when required.

Under similar conditions some authorities speak highly of the continued use of small doses of *iodide of potassium* (Bartels) or *bichloride of mercury*.

What are the conditions under which the patient should be sent to bed?—Whenever there is dropsy or the urine diminishes, or the albumen is greatly increased, or any inflammatory complication comes on, or there is marked dyspnœa, or any sign of uræmia, it is prudent to send the patient to bed, as it is better to err by being a little over careful.

The treatment of symptoms should not be undertaken too vigorously. Minor ones that are not causing distress are often best let alone, unless they are regarded as danger signals, when a few days rest in bed on simple diet and treatment will afford opportunity for judging of their value and of what is needed to be done. We should recognise the incurable nature of certain symptoms and avoid treating them. Of two remedies or methods of treatment we should choose that which is the less debilitating.

ALBUMINURIA.—In the last edition of his *Urinary and Renal Diseases*, Roberts repeats the question: "Are there any medicinal substances capable of exercising control over the quantity of albumen lost by the urine?" and he answers it as before, by saying: "Exact observations do not give an affirmative answer to this question."

Rosenstein deals with this question at length, and mentions a number of drugs tested by his assistant, Dr. Kooi and himself. He gives his opinion in the following terms: "The physician must be impressed with the idea that we possess no drug which can in any sense act upon the local disorder so as to diminish decidedly the excretion of albumen."

The action of drugs on diseased processes in the human organism is always an obscure and difficult problem, on account of the large number and inconstancy of the factors that enter into it. An exact observation demands identity of all conditions except the one to be tested, namely, the drug employed; but that is impossible, for no two cases are identical, nor is the same patient this week in identically the same circumstances as he was a week ago, or will be a week hence; so we must always allow an element of uncertainty in our most carefully devised experiments with drugs in disease, and our conclusions must be based rather on wide experience than upon the minute observation of particular cases.

Nevertheless, I have endeavoured to combine these two methods; I have made observations upon many cases, and have sought in the practice of my colleagues to learn the results of their treatment, so that by comparison I might eliminate the "personal equation" from my conclusions.

It is obviously essential to the proper treatment of this question that we should understand the natural course of the disease or symptom which we seek to influence with drugs. It is to my mind beyond question, and, if necessary, I could quote cases to prove it, that in acute Bright's disease and even in subacute Bright's disease, the albumen may diminish and disappear without the use of any drugs whatever. Moreover, cases of chronic Bright's disease are liable to intercurrent acute exacerbations of greater or less intensity, which may clear up just like an acute attack in healthy organs, the subsidence being accompanied by a more or less rapid and considerable reduction in the amount of albumen excreted. Finally, cases of persistent albuminuria present fluctuations in amount which depend upon causes that escape our present powers of observation.

Besides distinguishing between these phases of disease and allowing as best we can for these fluctuations, we have to bear in mind that the amount of albumen is influenced by diet, increased by exercise, and diminished by the recumbent position.

The subjects of Bright's disease are frequently also the subjects of other accidental or secondary organic diseases, such as heart, lung, or liver diseases, which probably play some part in the production of albuminuria, and interfere with the action of the drugs we employ, or give rise to fluctuations which depend upon changes originating in altered states of those organs.

A more obvious source of fallacy, but one which is certainly apt to be overlooked in private practice is, that the amount of albumen excreted does not always hold the same relation to the quantity of water, so that diuresis, either spontaneous or the effect of drugs, often produces an apparent improvement in the albuminuria, though, in fact, it may remain the same, or be actually increased in amount. Another and less excusable error may arise from comparing samples of urine passed at different periods of the day, as it is well known that the albumen varies very much, being generally most in the forenoon, other conditions being equal. No comparison should be made for the purposes of exact observation, except between samples of the whole twenty-four hours' urine collected and mixed together.

I carefully endeavoured to meet all the difficulties I have enumerated, and to eliminate these sources of fallacy.

The quantity of albumen was estimated by Esbach's tubes, generally by myself, always under my supervision; and I would bear testimony to the constancy of the results obtained by this method, which also gives approximately correct quantitative results. Before Esbach's tubes were introduced I used test-tubes marked with a scratch, and measured by means of a scale, which afforded a rough means of estimating the amount of albumen.

Perhaps no cases afford a better opportunity for testing the influence of drugs than those of "functional albuminuria," where the albumen is persistent, and is unaccompanied by any other evidence of disease. Many of my patients, otherwise apparently healthy, have been applicants for insurance, and both they and I have been anxious to remove what appeared to be the only obstacle to their acceptance. Yet I must admit that I have never been able to cure one of these insurance cases. After a few months of non-success, they have passed from under my observation, and I cannot tell what has become of them. Even those cases of "functional albuminuria" which I have seen ultimately cured by time, have not appeared to me to owe this result in any instance to the direct action of drugs.

Having made this general statement, I will submit my conclusions as to the actions of particular drugs a little more in detail.

Alkalies.—I have used alkalies in the form of diluents; for example, a quart of bitartrate of potash imperial (ʒss to Oj) daily, or a bottle of Vichy water (Célestins or Haute Rive); and in a series of chronic cases, with persistent and copious albuminuria, the results were distinctly favourable. In addition to these diluents, I have employed citrate of lithia, bicarbonate of potash, and benzoate and bicarbonate of soda, and include them in this favourable opinion. This effect was not due to the formation of alkali albumen, as Esbach's fluid precipitates this. In the following five chronic cases, the average total amount of albumen passed on the first and last two days of the experiment are given.

Name.	First Two Days.	Last Two Days.
B.	38 grs.	21 grs.
R.	38 grs.	13 grs.
J.	128 grs.	64 grs.
McD.	52 grs.	48 grs.
St.	16 grs.	21 grs.

Tannate of Soda.—I found the drug supplied under this name so nauseous that I have used in its stead the following formula.

℞ Acidi tannici, sodæ bicarb., āā gr. x; glycerini ♏xv; aq. ʒj. M. t. d. s. I can report in relatively favourable terms of it. For example:

Name.	First Two Days.	Last Two Days.
M.	198 grs.	150 grs.
J.	165 grs.	96 grs.

These were both chronic cases. In an acute case I got good results, but the influence of the drug was very doubtful here.

Name.	First Two Days.	Last Two Days.
Be.	22 grs.	11 grs.

Nitro-Glycerine.—I have seen cases do remarkably

well under the use of this drug; but I am not able to confirm this by exact observation, except in acute cases.

Name.	First Two Days.	Last Two Days.
Be.	26 grs.	10 grs.

In a chronic case I found:

Name.	First Two Days.	Last Two Days.
McD.	40 grs.	43 grs.

Fuchsin.—Under the most favourable circumstances I have been unable to observe results which bear out the reputation this drug has acquired.

Name.	First Two Days.	Last Two Days.
McD.	34 grs.	65 grs.
St.	66 grs.	74 grs.

I have used it in a very large number of cases, and I have never seen any good effected by it.

Digitalis appears to increase the amount of albumen, and this holds good of other heart-tonics, for example, *caffeine, strophanthus,* and *sulphate of sparteine. Iron,* including the acetate, sulphate, and perchloride, has the same effect of increasing the albumen. *Terpine,* in ten-grain doses, three times daily, in one case increased, in another did not diminish, the albumen. *Apocynum* increased the albumen in two cases, and diminished it in one. I was not able to observe the remarkable diuretic effect of this drug (used as the tincture in drachm doses) which is claimed for it across the Atlantic. I have used *turpentine* in several cases, without being convinced of any beneficial result, though hæmaturia has followed the employment of even minute doses (one minim). The *bichloride of mercury,* recommended by Millard, of New York, and in use by the homœopaths, has had a fair trial in the suggested doses (gr. $\frac{1}{1000}$), but has entirely failed. *Purgatives* and *diaphoretics,* though of great value in its treatment, do not appear directly to influence the amount of albumen excreted in chronic Bright's disease.

The following is a list of drugs whose action on albuminuria was tested in these experiments. Bitartrate of potash, bicarbonate of potash, citrate of lithia, carbonate of lithia, bicarbonate of soda, benzoate of soda, tannate of soda, tannic acid, digitalis, scoparium, sulphate of sparteine, strophanthus, pilocarpine, Trousseau's diuretic wine, caffeine, apocynum, cannabinon, ergot, turpentine, terpine, copaiba, oil of sandal-wood, fuchsin, anti-hydropin (pulvis blattæ orientalis), cantharides, iodide of potassium, chloral, spirits of nitrous ether, perchloride of iron, sulphate of iron, acetate of iron, acetate of lead, tartrate of antimony, sulphate of alum, bichloride of mercury, elaterium, jalap, scammony, guaiacum, and sulphur.

HÆMATURIA.—Slight hæmorrhage from the kidneys is not of much importance; in acute congestion it probably does good, at any rate I believe I have noticed that acute cases in which some moderate hæmaturia has been very persistent have eventually done very well, but when treatment is desirable a *sinapism* may be applied to the loins, or the same region may be *dry-cupped*.

Hæmostatic drugs are not very efficient. In urgent cases 2 or 3 grains of *ergotin* may be injected subcutaneously; or half-drachm doses of liquid *extract of ergot* given by the mouth, hourly if necessary, as in those cases of profuse hæmaturia which sometimes occur. Other remedies are *acetate of lead*, 1 to 3 grains every four hours; *tincture of hamamelis*, 15 to 20 minims every two to four hours; *gallic acid*, 10 grains every four hours.

DROPSY.—The knowledge we possess of the pathology of dropsy enables us at least to see what are the desirable points towards which to direct our treatment.

In the dropsy of acute nephritis and its consequences we should endeavour to keep the quantity of fluid swallowed within moderate limits, and to promote its evacuation by the bowels by the use of purgatives causing watery stools, of which one of the best is bitartrate of potash. This may be given as an electuary composed of one part of the salt to two parts of honey or syrup, a tea-spoonful for a dose, repeated until the desired effect is produced. Although an acid salt, it does not diminish the alkalinity of the blood serum, which would be very undesirable; and we may endeavour to maintain this by giving farinaceous food, with Vichy water and lemon juice as a drink.

Compound jalap or compound scammony powders or the resin of scammony may be used, but very drastic purgatives, such as elaterium, are not desirable.

Hot air baths should be employed daily.

There need be no hurry to employ drainage, unless urgent symptoms arise; if there be much fluid in the serous sacs this can be drawn off by a fine needle with tubing attached, but the cellular tissue is less easy to drain; cellulitis is sometimes set up by the attempt, and I have seen the effusion apparently poured out as fast as it was drawn off by the small trocars of Southey, so that although there was a considerable amount of fluid withdrawn the œdema did not lessen. If it has to be done, these small trocars with tubing attached are to be preferred to incisions, however small.

Œdema of the *prepuce* and *scrotum* may be relieved by several *punctures* with a darning needle; the parts afterwards being wrapped in absorbent cotton wool.

Œdema of the glottis is treated by *ice* to suck, *puncture*, and in the last resort *tracheotomy*. It is possible that *intubation* of the larynx might be effectual in relieving this symptom.

The action of the kidneys should be promoted by hot poultices to the loins, by diuretics, such as acetate of potash, squill, caffeine, digitalis, triticum repens, broom-tops, Trousseau's diuretic wine, &c., but diuretics are very uncertain remedies.

In the dropsy of chronic lithæmic nephritis we have to deal with a *failing heart*, and our chief reliance must be upon drugs that increase its energy. Digitalis takes the first rank; if there is aortic regurgitation convallaria, in 15 minim doses of the tincture, is a good drug; either may be combined with a grain of citrate of caffeine, 5 minims of liquor strychninæ, one minim of nitro-glycerine solution (1 per cent.), and an ounce of infusion of broom-tops.

If there is peritonæal effusion this may be removed by tapping. Pleural effusions should not be rashly interfered with, especially if there is reason to believe they are not recent.

These cases are eminently unsuited for purging; it is sufficient to keep the bowels open; for it must be remembered that when dropsy sets in from heart failure, the end is very near, and may be more easily precipitated than averted.

Uræmia.—*Headache* and *giddiness* are relieved best by citrate of *caffeine*, 1 to 2 grains in pill, or *nitro-glycerine*, 1 to 2 minims of the 1 per cent. solution in a little water. To ward off uræmic attacks *sodium benzoate* may be given. Thomson reports favourably of *benzoic acid*.

All the minor phenomena of uræmia—*blindness, deafness, cramps, formication, numbness* and *palpitation*, should be treated by a smart *purgative* and the *hot air bath*, or the injection of $\frac{1}{32}$th of a grain of *pilocarpine* under the skin. *Palpitation* may be relieved by nitro-glycerine.

Convulsions if soon over scarcely call for treatment; they may be stopped by *chloroform* inhalation or by *chloral* (gr. xx) injected into the rectum, and small doses (gr. v to x) of the same drug may be given every three or four hours to prevent their return; the skin should be got to act by the *hot air bath* or *pilocarpine*, and the *bowels* freely moved.

In obstinate convulsions *12 to 16 ounces of blood* may be taken from the arm.

Hemiplegia and *coma* call for bleeding to the like amount, but the *hot air bath* recovered one patient from coma after the attempt to bleed had failed.

It has been suggested to treat uræmia by *dilution of the blood* as has been practised in diabetic coma, and a case has been quoted in which temporary improvement followed the adoption of this plan.

Vomiting and *diarrhœa* are natural efforts at elimination which should not be too quickly checked.

Uræmic asthma is not at all amenable to treatment. Carter, of Liverpool, recommends drachm doses of *ozonic ether* in liq. ammoniæ acetatis, but it has not proved of any service in my hands. Inhalations of *oxygen* have been recommended, but are not ordinarily available. The ordinary treatment for uræmia may be tried. *Nitroglycerine* or *nitrite of amyl* affords some relief. *Ethylnitrite* may be tried in half-drachm doses of a 3 per cent. solution (LEECH). Roosevelt has recommended *cobaltonitrite* of potassium in half-grain doses every two to four hours.

EPISTAXIS and other external hæmorrhages are best treated by *ergotin*, two to three grains, injected under the skin, or *ergot* may be given by the mouth. An *ice-bag* should be applied locally.

INFLAMMATORY COMPLICATIONS — *pneumonia, pleurisy, pericarditis* and *peritonitis*, are of most unfavourable prognosis. Very active treatment is hardly judicious. *Poultices* or *hot fomentations* should be applied locally, and the case managed on general conservative principles.

GASTRIC CATARRH.—This symptom is often very troublesome. Green vegetables and fruits should be temporarily eliminated from the diet; a tea-spoonful of *Carlsbad salt* dissolved in a tumbler of hot water should be sipped each morning before breakfast, and the following powder given in a little milk shortly before each meal :

℞ Bismuthi Carb. gr. x
Sodii Bicarb. gr. x
Pulv. Rhei gr. ij
Pulv. Cinnamomi Co. gr. v
Fiat pulvis.

The liver should be gently stimulated by a pill containing *euonymin*, such as that given on p. 272, taken before the principal meal of the day or at bedtime.

BIBLIOGRAPHY.

BARTELS (C.). *Op. cit.*, p. 490.

CARTER (W.). Clinical Reports on Renal and Urinary Diseases. London, 1878, pp. 149, 150.

LEECH (D. J.). Nitrite of Ethyl. "Med. Chronicle," Vol. IX., 1888, p. 177.

MILLARD (H. B.). A Treatise on Bright's disease of the Kidneys. Second edition. New York, 1886, p. 229.

ROBERTS (W.). *Op. cit.* Fourth edition, p. 491.

ROOSEVELT. On Cobalto-Nitrite of Potassium. "New York Med. Jour.," Aug. 25th, 1888. Quoted in "The Practitioner,' Vol. XLI., p. 457.

ROSENSTEIN (S.). *Op. cit.* Third edition, p. 345.

THOMSON (R. STEVENSON). *Op. cit.*, p. 18.

GENERAL INDEX.

	PAGE
ANDERSON, *pp.* 90, 118; Andral, 30; Aufrecht, 41, 173	
Aceto-acetic Acid	122
Acetone	122
— Test for	122
Acid Blood	78
Acute Diseases, as causes of Nephritis	172
Acute Febrile Nephritis	173
Age, influence of	156
Albumen, diffusibility of	12
— Estimation of	129
— in Urine	125
— Tests for	126
Albuminuria	1, 125
— Causes of	5
— Cyclical	9
— Food-	13
— in Acute Febrile Nephritis	177
— in Chronic Febrile Nephritis	191
— in Lithæmic Nephritis	211
— in Obstructive Nephritis	234
— in Dyspepsia	10
— in Healthy Persons	7
— Mechanism of	11
— of Adolescence	9
— Remittent	9
— Theories of	12
— Treatment of	278
Albuminuric Retinitis	81, 215
Alcapton	118
Alcohol, as cause	159, 202
— in treatment	272
Algiers	269
Alimentary System, derangements of, in Chronic Bright's Disease	267
Alkali Albumen	1
Ammonia	161

	PAGE
Amnesia	250
Amyloid degeneration	149
Anatomy of Kidney	164
Animal Food	203
Anuria	98
Arsenic	161
Arterio-capillary Fibrosis	53
Ascending Nephritis	231
Asthma, Uræmic	247
— Treatment of	281
BAGINSKY, *p.* 122; Bamberger, 19, 50, 53, 150, 173; Barling, 237; Barreswil, 13; Bartels, 24, 27, 37, 45, 50, 64, 74, 110, 149, 272; Beale, 117; Béchamp, 142; Belfanti (Mya and) 3, 142; Berlioz (Yvon and) 98, 109; Bernard, 20; Biermer, 75; Blackall, 147; Bock, 31; Bostock, 54; Bouchard, 2, 77, 78, 140, 173; Bouillaud, 53; Brailey, 88; Breusing, 142; Bright, 11, 49, 58, 75, 147; Broadbent, 59; Brown-Séquard, 13; Brücke, 14; Brunton (Lauder), 13, 35, 128; Buhl, 53, 58; Bull, 122	
Bathing	5, 270
Bichloride of Mercury	272
Bile Acids, test for	123
Bile Pigment in Urine	122
— Test for	123
Bilirubin in Urine	122
Biliverdin in Urine	123
Bladder Diseases	161, 231
Blindness, Uræmic	249
Blood Casts	37

	PAGE
Blood in Urine	130
— Tests for	134
Blood-vessels of Kidney	168
Bournemouth	270
Bowman's Capsule	166
Braemar	270
Brain, Œdema of	74
Bronchitis	263
Buxton	270

CAMERER, p. 109; Carter, (A. H.), 14; Carter (W.), 74, 281; Charcot, 20, 149; Charrin and Bouchard, 173; Chateaubourg, 6; Christison, 13, 30, 54, 64, 74, 75; Cohnheim, 27, 29, 34; Colasanti, 116; Cook, 108, 113; Crooke, 184, 237; Crookshank, 237

Calculus	161
Cantharides	160
Carbolic Acid	161
Carbonate of Ammonia in Uræmia	76
— of Lime in Urine	107
Cardio-vascular Changes	45
Casts in Urine, 37, 137, 177,	191, 211
Catheterism	161, 231
Causation of Bright's Disease	153, 172
Cerebral Hæmorrhage	257
Cheyne-Stokes Breathing	247
Chlorides in Urine	104
Cholesterin in Urine	125
Choroidal Hæmorrhage	89
Chronic Bright's Disease, Complications of	243
Chronic Diseases, as causes of Nephritis	172
Chyluria	124
Classification	151
Climate	155, 200, 269
Clothing	270
Cobalt	161
Cobalto-Nitrite of Potassium	281
Collecting Tubes	167
Coma	253
Convoluted Tubules	166

	PAGE
Convulsions	253
Cramps	252
Cystin in Urine	117
Cystitis	161, 236, 239

DICKINSON, p. 45; Dobrowolsky, 81; Donders, 27; Doremus, 111

Deafness, Uræmic	249
Delirium	249
Diacetic Acid	122
Diet, influence of	159
— in Treatment	185, 196, 270
Digitalis	272
Distal Convoluted Tubule	167
Dropsy	23
— Definition of	24
— Fluid of	24
— in Acute Febrile Nephritis	173
— in Chronic Febrile Nephritis	190
— in Lithæmic Nephritis	219
— in Obstructive Nephritis	235
— Prevalence of	23
— Theories of	25
— Treatment of	278
Duration of Bright's Disease	179, 192, 221
Dyspepsia	203
Dyspnœa, Uræmic, 228, 247,	264, 266

EALES, pp. 10, 82, 88, 91, 92, 93; Ebstein, 119; Edlefsen, 24; Edmunds, 88; Ehrlich, 105; Eichhorst, 51; Esbach, 129; Ewald, 46, 52, 53, 59

Egypt	269
Endarteritis	52
Engadine	270
Epistaxis	208, 281
Epithelial Casts	89
Etiology of Bright's Disease in general	153
— of Acute Febrile Nephritis	172
— of Chronic Febrile Nephritis	187
— of Lithæmic Nephritis	199

INDEX.

	PAGE
Etiology of Obstructive Nephritis	231
Euonymin... 272,	282
Exciting Causes of Bright's Disease	159
Exercise	270

FEITELBERG, *p*. 35; Feltz, 76; Féré, 235; Flugge, 109; Fraenkel, 141; Francotti 124; Frerichs, 23, 30, 75, 76, 148; Friedländer, 141; Frost, 124; Fürbringer, 152

Fat in Urine	123
Febrile Diseases, as causes of Bright's Disease	159
— Nephritis	172
— Urine	178
Fehling's Solution, formula for	120
Ferrein, pyramid of	165

GALABIN, *pp.* 47, 58; Galezowski, 82, 83; Garrod, 155, 201; Gaucher, 160; Gauthier, 76; Gavarret, 30; Gee, 106; Germont, 41, 231, 233; Gibson, 245; Goodsir, 170; Gowers, 88, 89, 91; Graham, 15; Gram, 142; Grawitz, 47, 56, 59, 150; Gregory, 54; Gréhant, 75; Grützner, 3, 51, 55, 66, 67; Gull, 2; Gull and Sutton, 46, 51, 149

Gastric Catarrh	281
Giddiness	248
Gout	201
Gouty Kidney	201

HALES, *p.* 26; Hamilton, 59; Hanot, 47; Hare (Woodhead and) 142; Harris, 128; Hay, 123; Heidenhain, 55, 67; Heller, 127; Herman, 2, 27, 127; Hoffmann, 31; Holvotschiner, 142; Hoppe-Seyler, 5, 76; Huschke, 169

	PAGE
Hæmatemesis	268
Hæmaturia	130
— Causes of ...131, 177,	211
— Treatment of	278
Hæmoglobinuria	132
Hæmorrhage from Bowel	268
— into Vitreous	89
Hæmorrhoids	268
Hard Water 202,	271
Headache	249
Heart Disease as Cause	203
— Hypertrophy of, 45, 176, 190,	235
— in Acute Febrile Nephritis	176
Hemialbumose	3
Hemiopia	249
Hemiplegia, Transient	249
Henle's Loop	167
Heredity 157,	200
Hiccough in Uræmia	245
Hippuric Acid	115
History of Bright's Disease	147
Hot-air Bath 186, 196,	270
Hyaline Casts 39,	137
Hydræmic Dropsy	30
Hydrochinon in Urine	118
Hydrochloric Acid	161
Hypoxanthin in Urine	115

ISRAEL, *pp.* 47, 56, 59, 150

Indican in Urine	118
Inflammatory Complications	281
Interlobular Arteries	168
— Veins	168
Iodoform	161
Iron, as cause	161
— in treatment	277
Irregular Tubule	167

JACCOUD, *p.* 26; Jackson (Hughlings), 86, 91; Jean, 235; Johnson, 47, 50, 55, 65, 67, 148, 149, 153; Jones (Bence) 3, 5

Jaeger System	270

KANNENBERG, *p.*139; Kelsch, 42, 43, 149; Kierulf, 27; Kirk, 119; Klein, 165

286   BRIGHT'S DISEASE.

	PAGE
Kidney, Anatomy of	164
— Connective Tissue of	170
— Fœtal	165
— Lymphatics of	170
— Nerves of	170
— Shape of	164
— Vessels of	168
— Weight of	164, 171
Kreatinin	115

LAENNEC, p. 148; Landois, 35, 67, 108; Landois and Stirling, 113; Langhans, 40; Lassar, 5; Lecorché, 106; Legg, 132; Leichtenstern, 54; Le Nobel, 122; Lépine, 2, 12, 16, 19, 77; Leube, 7, 8; Leube (Salkowski and), 40, 142; Lichtheim, 29; Liebermeister, 75; Litten, 81, 173; Löbisch, 20; Ludwig, 17, 50; Lustgarten, 139

Labyrinth	165
Lactic Acid in Urine	116
Lardaceous Degeneration	149, 152, 192
Lead	161, 202
Leucin in Urine	117
Lithæmic Nephritis	151, 199
— Treatment of	269
Liver, in Chronic Bright's Disease	268
Lymphatics of Kidney	170

MACDONALD, p. 233; Mac Munn, 15; Magendie, 26; Mahomed, 58, 59, 153, 200, 203; Mannaberg, 139; Manson, 124; Marcacci, 8; Masterman, 271; Miley, 91; Mircoli, 173, Mobius, 43; Moscatelli, 116; Moxon, 9, 11, 149; Müller, 119; Murchison, 55, 160; Mya and Belfanti, 3, 142

Malpighian Bodies	167
— Pyramids	164
Manganese	161

	PAGE
Medullary Rays	165
Mercury, as Cause of Nephritis	161
Mercuric Chloride, in Treatment	272
Micro-organisms in Nephritis	172
— in Urine	139
Mucin in Urine	130
Myocarditis	53

NAUWERCK, pp. 104, 161; Newman, 66; Niemeyer, 26; Nussbaum, 18

Nephritis, Febrile	172
Nephritis, Acute Febrile	173
— Etiology	172
— Morbid Anatomy	173
— Dropsy	175
— Temperature	175
— Heart	176
— Pulse	176
— Ophthalmoscopic Appearances	176
— Urine	177
— Diagnosis	178
— Complications	178
— Duration	179
— Prognosis	179
— Illustrative Cases	180
— Treatment	185
Nephritis, Chronic Febrile	187
— Etiology	187
— Morbid Anatomy	187
— Dropsy	190
— Heart	190
— Pulse	191
— Ophthalmoscopic appearances	191
— Urine	191
— Diagnosis	191
— Complications	191
— Duration	192
— Prognosis	192
— Illustrative Cases	192
Nephritis, Lithæmic	199
— Etiology	199
— Morbid Anatomy	204
— Blood	212
— Dropsy	219
— Heart	212

INDEX.

	PAGE
Nephritis, Pulse	214
— Ophthalmoscopic Appearances	215
— Saliva	212
— Urine	209
— Diagnosis	219
— Complications	243
— Prognosis	221
— Illustrative Cases	223
— Treatment	269
Nephritis, Obstructive	231
— Etiology	231
— Morbid Anatomy	232
— Dropsy	235
— Heart	235
— Urine	234
— Uræmia	235
— Diagnosis	235
— Prognosis	236
— Illustrative Cases	236
— Treatment	241
Neuritis	85
Neuro-retinitis	85
Nickel	161
Nitric Acid	161
OLLIVIER, *pp.* 55, 202; Oppler, 76; Ord, 119	
Obstructive Causes of Bright's Disease	161, 231
Obstructive Nephritis	231
Œdema of Genitals	279
— of Glottis	261, 279
— of Lung	263
Oliguria	98
Ophthalmoscopic Appearances in Acute Febrile Nephritis	176
— in Chronic Febrile Nephritis	191
— in Lithæmic Nephritis	215
Opium	161
Organisms in Nephritis	172
Ovarian Tumour in Obstructive Nephritis	161
Oxalate of Lime in Urine	116
Oxalic Acid	161
— in Urine	116
Oxaluric Acid in Urine	116
Oxygen	281
Ozonic Ether	281

	PAGE
PARKES, *pp.* 13, 15; Paton (Noel), 113; Pavy, 9, 14, 15, 16; Peabody, 75; Piorry, 74; Pöhl, 125; Poncet, 89; Posner, 4, 6, 17, 18, 125; Potain, 55; Prout, 107, 108; Purslow, 72	
Pachymeningitis	260
Par-albumen	3
Pathological History of Bright's Disease	147
Pelvic Inflammation	161
Pemphigus	258
Peptone	2, 130
Peptonuria	130
Phosphates in Urine	105
Picric Acid as a Test for Albumen	128
Pneumonia	258, 268
Poisons, influence of	160
Polyuria	64, 98, 209
Prevalence of Bright's Disease	153
Previous Diseases, influence of	159
Propeptone	3
Prostatic Enlargement	161
Protocatechuic Acid in Urine	118, 231
Proximal Convoluted Tubule	166
Puerperal Kidney	231
Pulse in Acute Febrile Nephritis	176
— Tension	47
— Tracing	48, 176, 191
— in Chronic Febrile Nephritis	191
— in Lithæmic Nephritis	215
— in Obstructive Nephritis	235
— in Uræmia	254
Purpura Hæmorrhagica, case of	258
Pyrocatechin in Urine	118
Pyuria	136
— Tests for	137
QUINQUAUD, *p.* 75	

RALFE, pp. 2, 106, 109, 150; Ranke, 110; Rayer, 54, 148; Rees, 54, 73, 75; Rehder, 28; Rendu, 55; Ribbert, 17, 19, 41; Riess, 69; Rilliet, 73; Ritter, 76; Roberts (W.) 69, 100, 102, 103, 114, 117, 120, 128, 140, 149; Robinson, 19; Rokitansky, 20, 149; Rommelaere, 76; Rosenstein, 23, 73, 75, 140, 149, 153; Rossbach, 124; Rubner, 109; Runeberg, 19
Registrar-General's Returns 154
Renal Epithelium in Urine 139
Renal Vein 169
Retina, Detachment of ... 89
Retinal Changes 81, 176, 191, 229
— Cases Illustrating ...83, 92
— Cure of 94
— Diagnostic and Prognostic value of ... 91
— Hæmorrhages 86, 217
— Vessels, disease of ... 88
Retractile Albumen ... 2

SALKOWSKI, p. 3; Salkowski and Leube, 40, 97, 115; Schiff, 20; Schmidt, 24; Schottin, 76; Schultze, 106; Schultzen, 110; Seegen, 120; Semmola, 12, 55, 155; Senator, 2, 5, 8, 18, 19, 20, 47, 49, 235; Sibson, 47; Smith (W. G.), 101, 105, 119, 177; Snyers, 75, 76, 77, 151, 155; Stewart (Grainger), 8, 13, 46, 59, 149; Stirling, 8, 128; Strauss, 41, 231, 233; Sutton (Gull and), 46, 51
Salutaris Water 271
Sarcin in Urine 115
Sarcinæ in Urine... ... 140
Serum-Albumen 1, 125
— Globulin 1, 125
Sex, Influence of... ... 156

Social State, Influence of 158
Sodium Benzoate 272, 280
South Africa 269
Spectroscopic Examination of Blood in Urine ... 136
Spiral Tubule 166
St. Leonards 270
Stomach Catarrh 267, 281
Succinic Acid in Urine ... 115
Sugar in Urine 119
— Causes 120
— Estimation 121
— Test 120
Sulphuric Acid 161
— in Urine 108
Surgical Kidney 231
Syntonin 1

TAIT (Lawson) p. 35; Teissier, 106; Thoma, 19, 51, 67; Thormählen, 4; Thudichum, 104; Tizzoni, 12, 155; Todd, 55, 201; Traube, 30, 47, 50, 53, 58, 73; Treitz, 76; Trousseau, 29; Türck, 81
Temperature in Acute Febrile Nephritis ... 175
Tonic Spasm 252
Torquay 270
Toxic Urine 78
Treatment of Acute Febrile Nephritis ... 185
— of Chronic Febrile Nephritis 195
— of Lithæmic Nephritis 269
— of Obstructive Nephritis 231
— Albuminuria 278
— Epistaxis 281
— Hæmaturia 278
— Inflammatory Complications 281
— Uræmia 280
— Asthma 281
— Dropsy... 278
— Gastric Catarrh ... 281
Tremor 252
Tube Casts in Urine 87, 187
Tubules, Arrangement of 166

INDEX.

	PAGE
Tubules, Diagram of	166
Twitching in Uræmia	253
Tyrosin in Urine	117
USTIMOWITSCH *pp.* 55, 67	
Ulcerative Endocarditis	181
Uræmia	69, 228
— Types of	69
— in Acute Febrile Nephritis	179
— in Chronic Febrile Nephritis	191
— in Chronic Bright's Disease	244
— Coma	253
— Convulsions	253
— Dyspnœa	228, 247, 264, 266
— Diarrhœa	245
— Hiccough	245
— Headache	249, 253
— Ptomaines in	76
— Skin	244
— Symmetrical Gangrene	245
— Tremor	252
— Vomiting	245
— Illustrative Cases	69, 75, 181, 183, 245, 250, 254
— Treatment of	280
— Theories of	73
Urate of Ammonia	114
— of Soda	114
Urea, Detection of	111
— Effect of, taken in food	58
— Estimation	111
— in Lithæmic Nephritis	210
— in Obstructive Nephritis	235
— in Normal Urine	108
— in Urine of Bright's Disease	110
— in Urine of Acute Febrile Nephritis	177
— in Urine of Chronic Febrile Nephritis	191
— in Uræmia	75
Ureter	170
— Compression of	231
— Ligature of	231
Uric Acid, as Cause of Bright's Disease	160
— Causes	113

	PAGE
Uric Acid Crystals	113
— Detection	114
— Estimation	114
— in Lithæmic Nephritis	211
— in Normal Urine	113
Urine, Acetone in	122
— Acidity	102
— Albumen in	1, 125
— Alkalinity	102
— Bile in	100, 122
— Blood in	100, 130
— Chemical Composition	104
— Colour	99
— Influence of Drugs on Colour	100
— Darkening of	100
— Density	100
— Description	97
— Fat in	100, 123
— in Acute Febrile Nephritis	177
— in Chronic Febrile Nephritis	191
— in Lithæmic Nephritis	209
— in Obstructive Nephritis	234
— Mucin in	130
— Odour	99
— Peptone in	130
— Pus in	100, 136
— Quantity	99
— Reaction	102
— Sugar in	119
— Translucency	99
Uriniferous Tubules	165
— Diagram of	166
Urinometer	101
Urinometry	101
Urobilin	99
— Febrile	100
Uroleucic Acid in Urine	118
Uroxanthic Acid in Urine	118
Uterine Tumours	161, 231

VAN HELMONT, *p.* 147; Vierordt, 75; Virchow, 35, 148, 149, 150; Vogel, 106; Voit, 77; Vulpian, 12, 212

Vapour Bath	183, 196, 270
Vasa Recta	169

	PAGE		PAGE
Venæ Rectæ	170	Woodhead and Hare, 142	
— Stellatæ	169	Water, Hard	202, 271
Vichy Water	271	Waxy Degeneration	149, 152

WAGNER (E.), *pp.* 27, 82, 160; Webster, 92; Weigert, 150; Weil, 235; Wells, 81; Werner, 2; Weston, 39, 43; Willis, 108; Windle, 122; Wittich, 17;

Xanthin in Urine ... 115

YVON AND BERLIOZ, *pp.* 98, 109

Zinc 161

www.ingramcontent.com/pod-product-compliance
Lightning Source LLC
Chambersburg PA
CBHW031327230426
43670CB00006B/258